Lecture Notes in Networks and Systems

Volume 99

The series "Lecture Notes in Networks and Systems" publishes the latest developments in Networks and Systems—quickly, informally and with high quality. Original research reported in proceedings and post-proceedings represents the core of LNNS.

Volumes published in LNNS embrace all aspects and subfields of, as well as new challenges in, Networks and Systems.

The series contains proceedings and edited volumes in systems and networks, spanning the areas of Cyber-Physical Systems, Autonomous Systems, Sensor Networks, Control Systems, Energy Systems, Automotive Systems, Biological Systems, Vehicular Networking and Connected Vehicles, Aerospace Systems, Automation, Manufacturing, Smart Grids, Nonlinear Systems, Power Systems, Robotics, Social Systems, Economic Systems and other. Of particular value to both the contributors and the readership are the short publication timeframe and the world-wide distribution and exposure which enable both a wide and rapid dissemination of research output.

The series covers the theory, applications, and perspectives on the state of the art and future developments relevant to systems and networks, decision making, control, complex processes and related areas, as embedded in the fields of interdisciplinary and applied sciences, engineering, computer science, physics, economics, social, and life sciences, as well as the paradigms and methodologies behind them.

**** Indexing: The books of this series are submitted to ISI Proceedings, SCOPUS, Google Scholar and Springerlink ****

More information about this series at http://www.springer.com/series/15179

Hiren Kumar Deva Sarma ·
Bhaskar Bhuyan · Samarjeet Borah ·
Nitul Dutta
Editors

Trends in Communication, Cloud, and Big Data

Proceedings of 3rd National Conference on CCB, 2018

Springer

Editors
Hiren Kumar Deva Sarma
Department of Information Technology
Sikkim Manipal Institute of Technology
Majitar, Sikkim, India

Bhaskar Bhuyan
Department of Information Technology
Sikkim Manipal Institute of Technology
Majitar, Sikkim, India

Samarjeet Borah
Department of Computer Application
Sikkim Manipal Institute of Technology
Majitar, Sikkim, India

Nitul Dutta
Department of Computer Science
and Engineering
Marwadi University
Rajkot, Gujarat, India

ISSN 2367-3370 ISSN 2367-3389 (electronic)
Lecture Notes in Networks and Systems
ISBN 978-981-15-1623-8 ISBN 978-981-15-1624-5 (eBook)
https://doi.org/10.1007/978-981-15-1624-5

This Springer imprint is published by the registered company Springer Nature Singapore Pte Ltd.
The registered company address is: 152 Beach Road, #21-01/04 Gateway East, Singapore 189721, Singapore

Preface

Communication, cloud, and big data are three dominant technologies at the present moment. Due to the tremendous advancements of information and communication technology, electronics engineering, and computing science, the society has witnessed a paradigm shift in the approaches and the ways people communicate.

Cloud technology has become indispensable in the world of information technology. From storage of data to the burden of computing, people have started trusting the cloud. The role of a server computer in an organization is now played in the cloud.

Big data analytics has attracted the research community significantly in recent times, mainly due to its tremendous promise that it can offer for sophistication of human lives, as a whole. Due to the popularity of social networks and growing usage of Internet of Things (IoT) devices, there has been tremendous growth in data generation every day. In order to exploit the huge data available today, there is a need of having novel tools and techniques for processing these data, storage of these data, and communication of these data. At the same time, security to such data is a major concern today.

There are several issues to be addressed in the areas of communication, cloud, and big data in spite of the fact that people already have adopted these technologies along with Internet technology. The purpose of this book is to report latest research results in these areas along with research results in some allied topics related to computing.

The book contains 16 articles comprising five articles on communication, one on cloud, one on big data, and nine articles on various topics related to computation in general. These articles were presented during the National Conference on Communication, Cloud and Big Data (CCB) 2018, organized by Department of Information Technology at Sikkim Manipal Institute of Technology (SMIT), Sikkim, during November 2–3, 2018. CCB 2018 is the third in its series.

We are thankful to all the contributing authors of CCB 2018. We thank Sikkim Manipal Institute of Technology for extending all support in organizing CCB 2018. We are highly grateful to Prof. Janusz Kacprzyk for his guidance and encouragement throughout the journey. We thank the editorial team of Springer Nature for extending all support to us. We thank Springer Nature for supporting us as publication partner, all the way in the process of publishing this book.

Gangtok, India — Dr. Hiren Kumar Deva Sarma, Editor

Gangtok, India — Dr. Bhaskar Bhuyan, Co-editor

Gangtok, India — Dr. Samarjeet Borah, Co-editor

Rajkot, India — Dr. Nitul Dutta, Co-editor

Contents

Computation

Editors and Contributors

About the Editors

Dr. Hiren Kumar Deva Sarma is currently a Professor in the Department of Information Technology, Sikkim Manipal Institute of Technology, Sikkim, India. He received his B.E. in Mechanical Engineering from Assam Engineering College (University of Gauhati) in 1998, his M.Tech in Information Technology from Tezpur University in 2000, and his Ph.D. from Jadavpur University (Department of Computer Science and Engineering) in 2013. He has received the Young Scientist Award from the International Union of Radio Science (URSI) in the XVIII General Assembly 2005, held at New Delhi, India. He has published more than seventy research papers in various international journals and leading refereed international and national conferences. His current research interests are wireless sensor networks, mobility management in IPv6-based networks, cognitive radio networks, network security, robotics, distributed computing, and big data analytics.

Dr. Bhaskar Bhuyan is an Associate Professor in the Department of Information Technology, Sikkim Manipal Institute of Technology, Sikkim, India. He received his B.E. (1997) from Motilal Nehru National Institute of Technology, Allahabad, in Computer Science & Engineering, and his M.Tech. (2000) and Ph.D. (2017) from Tezpur University, Assam, India. His research interests include computer networks and distributed systems, and he has published a number of papers in these areas.

Dr. Samarjeet Borah is currently a Professor in the Department of Computer Applications, Sikkim Manipal Institute of Technology (SMIT), Sikkim, India. Dr. Borah is involved in various academic, research, and academic administrative activities, as well as curriculum development activities, and is a member of the board of studies, doctoral research committee, and IT infrastructure management at SMIT. Dr. Borah is Principal Investigator/Co-Principal Investigator for three projects sponsored by AICTE (Government of India), DST-CSRI (Government of India), and Dr. TMA Pai Endowment Fund. He is associated with IEEE, ACM (CSTA), IAENG, and IACSIT and has organized various national and international conferences at SMIT, such as an ISRO-sponsored training programme on Remote Sensing & GIS, NCWBCB 2014, NER-WNLP 2014, IC3-2016, ICACCP 2017, and IC3-2018. Dr. Borah is an editor/guest editor for various journals, like IJSE, IJHISI, IJGHPC, IJIM, IJVCSN, JISys, and IJIPT.

Dr. Nitul Dutta is currently a Professor in Department of Computer Science and Engineering, Marwadi University, Rajkot, Gujarat. He received his B.E. degree in Computer Science and Engineering from Jorhat Engineering College, Assam (1995), and M.Tech. in Information Technology from Tezpur University, Assam (2002). He received his Ph.D. (area of study: mobile IPv6) from Jadavpur University (2012). He has published more than thirty research papers in various leading journals and conferences. His research interests are mobile IPv6, cognitive radio networks, and distributed computing.

Contributors

A. Anand Sikkim Manipal Institute of Technology, Sikkim Manipal University, Majitar, Rangpo, Sikkim, India

Priyadarshini Arambam Amity Institute of Public Health, Amity University, Noida, Uttar Pradesh, India;
Batra Hospital & Medical Research Centre, New Delhi, India

Pradosh Bandyopadhyay Department of MCA, Techno India, Salt Lake, Kolkata, West Bengal, India

Monalisa Baneree Department of MCA, Techno India, Salt Lake, Kolkata, West Bengal, India

Sangeeta Banik Department of CSE, SNU, Kolkata, West Bengal, India

S. Bera Sikkim Manipal Institute of Technology, Sikkim Manipal University, Majitar, Rangpo, Sikkim, India

Sonam Lhamu Bhutia Sikkim Manipal Institute of Technology, Sikkim Manipal University, Majhitar, Sikkim, India

Rupanka Bhuyan ICFAI University Nagaland, Dimapur, India

Sangappa Ramachandra Biradar SDM College of Engineering and Technology, Dharwad, Karnataka, India

Samarjeet Borah Sikkim Manipal Institute of Technology, Sikkim Manipal University, Majhitar, Sikkim, India

Sudipta Chakrabarty Department of MCA, Techno India, Salt Lake, Kolkata, West Bengal, India

Ankan Mukherjee Das Amity Institute of Public Health, Amity University, Noida, Uttar Pradesh, India

Purnendu Das Department of Computer Science, Assam University, Silchar, India

Sanju Das Department of Computer Science, Assam University, Silchar, India

Ashis Datta Sikkim Manipal Institute of Technology, Majitar, Sikkim, India

Amlan Jyoti Dey NIIT Technologies Ltd., SEZ Greater Noida, Greater Noida, India

Nitul Dutta Faculty of Technical Sciences, Jan Wyzykowski University, Polkowice, Poland

M. K. Ghose Computer Science & Engineering, OmDayal Group of Institutions, Uluberia, Howrah, West Bengal, India

Sandip Ghoshal Department of MCA, Techno India, Salt Lake, Kolkata, West Bengal, India

Md Ruhul Islam Department of Computer Science and Engineering, Sikkim Manipal Institute of Technology, Sikkim Manipal University, Majhitar, Sikkim, India

Rishabh Jain Amity Institute of Public Health, Amity University, Noida, Uttar Pradesh, India

Rajiv Janardhanan Amity Institute of Public Health, Amity University, Noida, Uttar Pradesh, India;
Amity School of Public Health, Noida, Uttar Pradesh, India

Sachidananda S. Joshi SDM College of Engineering and Technology, Dharwad, Karnataka, India

Upendra Kaul Batra Hospital & Medical Research Centre, New Delhi, India

Archit Khatri Amity Institute of Public Health, Amity University, Noida, Uttar Pradesh, India

Himshekhar Nr. Konwar Department of Computer Science and Engineering, Sikkim Manipal Institute of Technology, Sikkim Manipal University, Majhitar, Sikkim, India

Kapil Kumar Department of Computer Science and Engineering, Sikkim Manipal Institute of Technology, Sikkim Manipal University, Majhitar, Sikkim, India

Nityam Mittal Amity Institute of Public Health, Amity University, Noida, Uttar Pradesh, India

Amiyangshu Nath Department of Computer Science and Engineering, Sikkim Manipal Institute of Technology, Sikkim Manipal University, Majhitar, Sikkim, India

Kushal Pokhrel Department of CSE, SMIT, Majitar, Sikkim, India

Ratika Pradhan Sikkim Manipal Institute of Technology, Sikkim Manipal University, Majhitar, Sikkim, India

Prativa Rai Department of Computer Science & Engineering, Sikkim Manipal Institute of Technology, Sikkim Manipal University, Majitar, Sikkim, India

S. Rai Sikkim Manipal Institute of Technology, Sikkim Manipal University, Majitar, Rangpo, Sikkim, India

Priya Ranjan Amity School of Engineering & Technology, Noida, Uttar Pradesh, India;
Amity Institute of Public Health, Amity University, Noida, Uttar Pradesh, India

Bishwa Ranjan Roy Department of Computer Science, Assam University, Silchar, India

Surojit Roy Department of MCA, Techno India, Salt Lake, Kolkata, West Bengal, India

Shazina Saeed Amity Institute of Public Health, Amity University, Noida, Uttar Pradesh, India

Mousam Saikia Sikkim Manipal Institute of Technology, Majitar, Sikkim, India

Hiren Kumar Deva Sarma Department of Information Technology, Sikkim Manipal Institute of Technology, Sikkim Manipal University, Majitar, Sikkim, India

Kumar Dron Shrivastav Amity Institute of Public Health, Amity University, Noida, Uttar Pradesh, India

Lolita Singh Marwadi University, Rajkot, India

S. N. Sur Sikkim Manipal Institute of Technology, Sikkim Manipal University, Majitar, Rangpo, Sikkim, India

Hariom Vashista Amity Institute of Public Health, Amity University, Noida, Uttar Pradesh, India

Communication

Routing Protocols for CRAHN: A Comparative Evaluation

Lolita Singh and Nitul Dutta

Abstract In wireless networks, the spectrum scarcity problem is migrated to much extent by cognitive radio network. Cognitive radio network has the capability to utilize the unutilized frequency spectrum. Due to change in data rates and channel availability, the task of routing in cognitive radio network has become a very challenging task. In this paper, we have given a review of different routing protocols that can be used in cognitive radio network before it we have given a brief description of cognitive radio network architecture that gives idea about things involved in cognitive radio network. We have also outlined various issues or challenges associated with routing in cognitive radio followed by different routing techniques. The last section of this paper gives brief idea about different approaches that can be used for reducing the effects of different routing issues in cognitive radio network. This paper helps in getting idea about cognitive radio network, their issues or routing protocols.

Keywords Cognitive radio · Cognitive radio network · Routing protocols · Routing issues

1 Introduction

In crowded ISM bands, there is an increase in congestion due to continuous increase in wireless devices. Mainly in current wireless networks, a fixed spectrum assignment policy is used by governmental agencies [1]. Over vast geographical areas, rights for various frequency band usages are granted by giving licenses. There is an increase in request for spectrum allocation authority that occurs by huge success in wireless applications. There is overcrowding due to sensor networks, mesh networks,

L. Singh (✉)
Marwadi University, Rajkot, India
e-mail: Lolita.singh101463@marwadiuniversity.ac.in

N. Dutta
Faculty of Technical Sciences, Jan Wyzykowski University, ul. Skalnikow 6 B, 59-101 Polkowice, Poland
e-mail: n.dutta@ujw.pl

H. K. D. Sarma et al. (eds.), *Trends in Communication, Cloud, and Big Data*, Lecture Notes in Networks and Systems 99, https://doi.org/10.1007/978-981-15-1624-5_1

WLANs, personal area networks, body area networks, etc. like wireless technologies operate in unlicensed bands. Wireless technology is the main factor in which the use of licensed band depends. Operator's commercial success and their market penetration are other factors to which there is assignment of frequencies. Current study done by the Federal Communications Commission (FCC) shows that there are only bounded geographical areas on which static assignment policies are used for very short period of time. Its average utilization is between 15 and 85% [2].

In order to solve the problem of spectrum scarcity, a cognitive radio (CR) comes into existence. Cognitive radio technology offers the solution as a disruptive technology innovation that will enable the future wireless world [3]. The spectrum efficiency gets improved by CR technology that utilized unused available spectrum efficiently. In order to improve the utilization of licensed frequency band that is assigned to licensed or PU can be used by cognitive radio networks (CRNs). The PUs are users who have license to access that band in which they are operating. All around the world, radio frequency (RF) uses as an important resource for safety, employment, communication and entertainment like different services. As compared to the problem of actual RF spectrum, physical availability spectrum usage is considered as a big problem addressed by FCC. Currently, used deployed policy of static spectrum allocation and RF band can be used by only licensed or PU is the main reason behind problem of spectrum unavailability. These findings need more efficient methods for utilization of the RF resources, and the cognitive radio (CR) technology is envisioned as a new mechanism for flexible usage of the RF spectrum [4]. The secondary users (SUs) or unlicensed users can operate in licensed bands using CRN technology or available RF spectrum channels can be utilized by them. Dynamic spectrum access (DSA) is SU ability to change its frequency of operation. The CRNs are defined as machine learning by Ryan W. Thomas and Daniel H. as by experience performance of CRN gets improved without having information of environment in which they are operating [5].

CRN has become an important research topic for different researchers as there are number of open issues than need to be solved. Most of the researches are done on different spectrum access approaches that can be used and on their accessibility that results in improved results in terms of power saving, etc. or on the lower layer (PHY and MAC) issues. Not much research is done on CRNs routing protocols and different issues associated with it. The spectrum utilization efficiency and its capability get improved using CR technology in wireless network. The complete paper is divided into different sections. The first section is an introduction about CR and the need of using it, and the second section gives architecture of CR. Different routing protocols are discussed in the third section, and later in the fourth section, we have given a brief description of different issues associated in it. In the last section of this paper, different approaches are discussed that can be used for removing different issues associated with routing protocols of CRN.

2 Architecture of CRN

In order to design novel protocol for communications, there is a need of clear description about cognitive radio network. Its architecture figure and detail of components are given in Fig. 1.

Infrastructural and Infrastructure-less CRN are two categories of CRN. The communication is done using CRN base station like fixed infrastructure components in case of infrastructural CRN. On the other hand, communication in case of infrastructure-less CRN is done without using fixed infrastructure or base station. This is similar to multi-ad-hoc network. The cognitive networks and primary elements are defined as follow:

Network that owns a separate RF spectrum band for TV broadcast networks and common cellular like its own services are known as **primary network**. Its components are given below:

The **primary user** or also known as licensed user has the right to operate in its RF spectrum band. Any interference by SU in PU RF spectrum band is not bearable by it. Another is **primary base station** or fixed infrastructure component that is used to controlling the licensed users' access. The sharing of RF spectrum band with CR users cannot be done by it.

There is no spectrum band in **cognitive network** that they can use for their communication. This is the reason it is also known as unlicensed networks. The **cognitive users** and **cognitive base station** are the components of cognitive network. The CU is known as secondary or unlicensed users because they do not have their own frequency band. They access the licensed band shared by users in an opportunistic manner, and there is a need of spectrum sensing like extra functionality to operate in licensed band. As the primary base station in cognitive capabilities, there is a

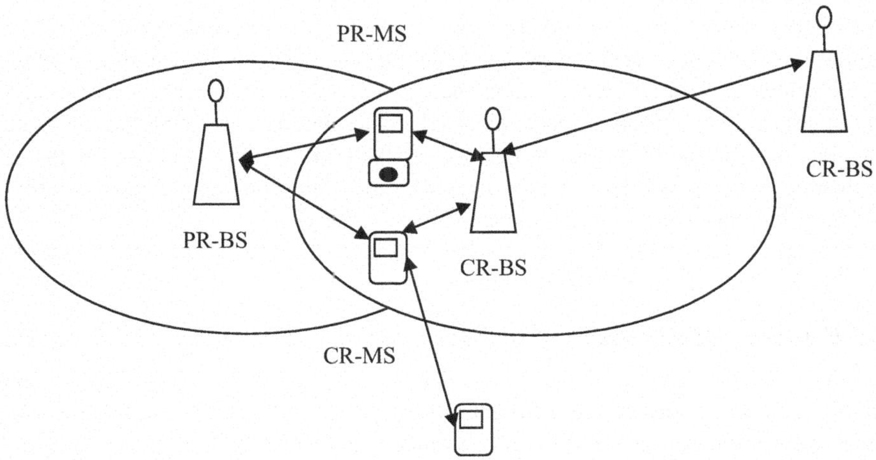

Fig. 1 Architecture of cognitive radio networks

fixed infrastructure component cognitive base station through which other network is accessed by secondary users. Single hop connection is provided to secondary users using it.

3 Routing in CRN

Next-hop node decision and decision of channel selection are the process of CRN routing. By the increase in number of technologies that utilize an unlicensed band and FCC provided statistics on licensed bands, underutilization leads to the need of routing in CRN [6]. There is a need of close coordination between spectrum manager and routing node due to the mobility of secondary users so that a proper routing decision can be taken [7]. Between secondary users' path failure can be occurred due to unexpected PU presence at a location that results in need of rerouting the path. In a network, the path is selected using a process of routing through which network traffic is sent. Or routing is also a process of sending data from source to destination. The routing process usually directs forwarding on the basis of routing table which maintains a record of the routes to various network destinations.

To get efficient routing, it is important to construct routing tables that are stored in router memory. At a time, only one network is used in most of the routing algorithms and multiple alternative paths use get enables by multipath routing techniques. In routing process, there is a change due to node and spectrum involvement in CRN routing. The movement of packet from one host to another is also known as process of routing. The wireless network routing is different from cognitive network routing [8].

As in case of cognitive radio network, a combination of spectrum management and traditional routing makes a routing process sin CRN in which there is variation is availability of network from node to node with respect to time and location [9]. The primary user activities get affected by availability of spectrum. In case of multi-hop CRNs routes are formed and maintained between SUs destination and source through intermediate nodes. Routing mechanism selects the intermediate node on the route and also decides the frequency to be used on each link of the route. So, routing in multi-hop ad hoc networks and mesh networks and multi-channel is same as routing in CRNs but have some extra challenges of PU activity.

4 Routing Protocols for CRNs

In this section, we have given different routing protocols for CRNs and along with their comparison in terms of different parameters associated with it.

Features	SORP [10]	LCB [11]	STOD-RP [12]	SAMER [13, pp 9–15]	SPEAR [14]	SEARCH [15]
Throughput (end to end)	Multi-flow and multi frequency scheduling	Delay is adaptive and cooperative with routing protocol	Better than scheme of hop count	High-throughput links are utilized for getting high throughput	Link base and flow base techniques are used	End-to-end latency is maintained and PUs avoided activity region
Route discovery	Broadcast RREQ messages	Same as SORP	Route announcement message (RANN) is broadcasted	Link state packets is used to broadcast	Control channel is used for RREQ broadcasting	RREQ messages broadcasted
Packet size of route discovery	Each node append SOP list	Each intermediate node append SOP list	Each and every node updates a field called cumulative metric	Depend on calculation of hop to hop	Time out field is used for route discovery	Based on geographic routing
Routing decisions	By MAC and network layer cooperation	Combined MAC and network layers is used for decisions	Not depend on cross-layer	By MAC and physical layer collaboration	Same as SPEAR	On the basis of geographic forwarding principle
Route nature selection of best path	On demand Path and node delay	On demand Decided by cumulative path delay	Periodical on the basis of global and local decision schemes	Periodical	On demand	On demand
Handling link failures	No	Yes, by flow redirecting to another node	–	No	Yes, time out field	–

SORP stands for spectrum-aware on-demand routing protocol, **CAODV** stands for cognitive ad hoc on-demand distance vector protocol, **STOD-RP** stands for spectrum-tree based on-demand routing protocol, **SAMER** stands for spectrum-Aware mesh routing in cognitive radio networks, **SPEAR** stands for high-throughput spectrum-aware routing protocol and **SEARCH** stands for spectrum-aware routing protocol.

5 Challenges and Issues of Routing Protocol in CRN

In this section, we will give review on different challenges related to CRN routing.

1. **Spectrum Availability or Link Availability**: Spectrum availability should be known by routing module that can be achieved using spectral environment monitoring. The available opportunities are fetched by DSA after sensing the RF band, so for communication, a licensed band is used by these networks. When there is less usage of frequency band by primary users, then these communication opportunities are available which shows that channel availability is based on time and geographic [16].
2. **Primary User Activity Awareness**: Delay, throughput, energy efficiency and throughput like measurement of route quality and primary user activities affect the topology of cognitive radio network. With spectrum availability, these things also need to be considered.
3. **Route Maintenance or Unidirectional Link**: There is a frequent route rerouting due to activities of primary user that result in network performance degradation. So, in case of convenient routing in CRN there is a need of effective procedures of signaling. There are more chances of unidirectional links in CRNs that are available for very short period of time rather than for minutes or hours. Due to it, there will be no guarantee that channel used for sending station will be available till that same channel is used for transmission by receiving station.
4. **Lack of CCC (Common Control Channel)**: Route establishment, neighbor and route discovery like some of the specific functionality are done using global or local broadcast messages in case of traditional routing protocols. Broadcast becomes a major problem in CRN due to lack of common control channel.
5. **Intermittent Connectivity**: There is frequent change in reachable nodes because of primary user and spectrum availability activities in CRN. So, network connectivity in cognitive radio network depends on spectrum availability.
6. **Deafness Problem**: In CRNs, availability of link is for short time due to link availability and channel diversity. The primary users' interference avoidance or route maintenance is done by switching between available channels due to deafness problem. This results in delay in CRNs communications [17].
7. **Heterogeneous Wireless Networks**: By heterogeneous and multiple wireless networks, CRN comes into existence. So, there is a need of routing because intersystem handover is critical, and channels or links are available

for an extremely short duration. Among these heterogeneous wireless networks, successful networking depends on cooperative relaying [18].

6 Different Approaches to Solve Various Issues Associated with CRN

In the below given table, we have given a summary of different solutions for routing. As given in the table, there are two main categories: One is based on static network topologies, and other is based on local radio resource management decisions.

Protocols	Performance
MINLP-MILP formulation based on graph [19]	It is good in whole path selection
Based on interference and power [20]	It is same as graph based
Based on delay [17, pp 1–6]	It is good in whole path and next-hop selection
Based on throughput [13, pp 1–5]	It is good in whole path selection, spectrum dynamic awareness and mobility support
Based on link quality/stability [21]	It is good in whole path and next-hop selection, reconfiguration to varying spectrum
Probabilistic approaches [22]	It is good in whole path selection, spectrum dynamic awareness and mobility support

On topologies of static network, proposals are focused on neighboring and spectrum occupancy and SUs' neighboring topological information. The network-state partial local radio resource management information is used in another proposal. This approach is also known as local spectrum-based knowledge approach. In case of first approach where graph theory and mathematical programming tools are used reduced the modeling/designing problems of CRNs to problem of classical wireless static network. A huge importance of this approach has been seen in different applications like the SUs have access to databases storing the spectrum maps even after the complex implementation and scarcely scalable. On the other hand, some approaches based on local information on spectrum occupancy gathered by each SU through local and distributed sensing mechanisms.

The whole path is set up in some cases, and on the other side, proposed approaches are based on the selection hop by hop of the next forwarding node. On each link of the path, spectrum selection is done by combining the distinguished characteristics of all routing approaches. Different metrics are used for this purpose that captures the available spectrum hole characteristics. Then, throughput, delay like QoS parameters and interference power like radio environment are used for appropriate spectrum band selections. In case of CRNs routing, data behavior of PU is considered as a key point. In fact, routes must explicitly provide a measure of protection to the ongoing communication of the PUs, while at the SUs, the side must guarantee stability when the PU behavior varies.

7 Conclusion

Cognitive radio networks (CRNs) can operate in the licensed frequency band to improve its utilization with the coexistence of the primary users (PUs) or licensed users. The PUs are the users who have license to access that band in which they are operating. Routing in CRN is a challenging task due to various issues and challenges involved in it. This issue occurs due to various reasons such as data rates, availability of channel and many more. This has a gain, a focus of number of researchers to introduce novel approaches to get rid of it. In this paper, firstly we have given a brief introduction about cognitive radio network forwarded with its architecture, that is, given brief detail about CR. Then, how routing in CRN takes place and what is its need in given in the third section. There are a number of routing protocols that can be used in CRN, and we have given a comparison of different techniques in terms of different parameters. The concept of routing in CRN comes with various issues and challenges in it. There is a need to discuss those issues in this paper we have given brief details of some of the existing issues and challenges that affects the operation of routing in CRN. The last section of this paper gives comparison and detail about different approaches that can be used for reducing the effects that affect routing operation in CRN. Still there is a need of lots of research in the field of designing CRNs routing which is a distinctive feature of CRNs.

References

1. Mitola III J (2001) Cognitive radio for flexible mobile multimedia. Mob Netw Appl 435–441
2. FCC and FCC (2002) ET Docket No. 02-135 November 2002 Federal Communications Commission. Communications 2
3. Akyildiz IF, Lee W-Y, Vuran MC, Mohanty S (2006) NeXt generation/dynamic spectrum access/cognitive radio wireless networks: a survey. Comput Netw 50:2127–2159
4. Dutta N, Deva Sarma HK (2015) A probability based stable routing for cognitive radio adhoc networks. Springer, pp 1–15
5. Thomas RW, DaSilva LA, MacKenzie AB (2005) Cognitive networks. In: Proceedings of IEEE DySPAN 2005, pp 352–360
6. Commission FC (2002) Spectrum policy task force. Technical report, Nov 2002
7. Cesana M, Cuomo F, Ekici E (2010) Routing in cognitive radio networks: challenges and solutions. Ad Hoc Netw
8. Subramanian AP, Gupta H, Das SR, Cao J (2008) Minimum interference channel assignment in multiradio wireless mesh networks. IEEE Trans Mob Comput 7:1459–1473
9. Dutta N, Deva Sarma HK, Polkowski Z (2017) Cluster based routing in cognitive radio adhoc networks: reconnoitering SINR and ETT impact on clustering. Elsevier, pp 1–13

10. Cheng G, Liu W, Li Y, Cheng W (2007) Spectrum aware on-demand routing in cognitive radio networks. In: Proceedings of the 2nd IEEE international symposium on new frontiers in dynamic spectrum access networks, Dublin, 17–20 Apr 2007
11. Yang Z, Cheng G, Liu W, Yuan W, Cheng W (2008) Local coordination based routing and spectrum assignment in multi-hop cognitive radio networks. Mob Netw Appl 67–81
12. Zhu GM, Akyildiz IF, Kuo GS (2008) STOD-RP: a spectrum-tree based on-demand routing protocol for multi-hop cognitive radio networks. In: GLOBECOM. IEEE Communications Society
13. Pefkianakis I, Wong S, Lu S (2008) SAMER: spectrum aware mesh routing in cognitive radio networks. In: 3rd IEEE symposium on new frontiers in dynamic spectrum access networks, DySPAN 2008
14. Sampath A, Yang L, Cao L, Zheng H, Zhao BY (2008) High throughput spectrum-aware routing for cognitive radio based ad-hoc networks. In: 3rd international conference on cognitive radio oriented wireless networks and communications, CrownCom 2008
15. Chowdhury KR, Felice MD (2009) Search: a routing protocol for mobile cognitive radio ad-hoc networks. Comput Commun
16. Cout D (2003) A high-throughput path metric for multihop wireless routing. In: Proceedings of MobiCom
17. Ma H, Zheng L, Ma X, Luo Y (2008) Spectrum aware routing for multi-hop cognitive radio networks with a single transceiver. In: 3rd international conference on cognitive radio oriented wireless networks and communications, CrownCom 2008
18. Cheng K, Prasad R (2009) Cognitive radio networks. Wiley
19. Wang Q, Zheng H (2006) Route and spectrum selection in dynamic spectrum networks. In: 3rd IEEE consumer communications and networking conference. CCNC 2006, vol 1, pp 625–629
20. Pyo CW, Hasegawa M (2007) Minimum weight routing based on a common link control radio for cognitive wireless ad hoc networks. In: IWCMC'07: proceedings of the 2007 international conference on wireless communications and mobile computing, pp 399–404
21. Filippini I, Ekici E, Cesana M (2009) Minimum maintenance cost routing in cognitive radio networks. In: IEEE 6th international conference on mobile adhoc and sensor systems. 2009, MASS'09, pp 284–293
22. Khalife H, Ahuja S, Malouch N, Krunz M (2008) Probabilistic path selection in opportunistic cognitive radio networks. In: IEEE global telecommunications conference, GLOBECOM 2008, pp 1–5

EERP: Energy-Efficient Routing Protocol Performance Analysis of DSR and AODV for Mobile Adhoc Network

Sachidananda S. Joshi and Sangappa Ramachandra Biradar

Abstract The multihop wireless network, especially mobile adhoc network, has gained popularity in various contexts that may include independent rescue system, connecting network at the edge in Internet and sub-network in upcoming global Internet of the cyber-physical system. The future-generation network systems will be a kind of software-defined network, where the system will have a possibility of choosing and adopting various routing protocol depending upon the context. This paper discusses the advantages of multihop wireless communication and investigates the performance behaviors of AODV and DSR with different Quality of Services parameters by proposing an enhanced energy model in the base node configuration.

Keywords Wireless communication · Multihop · MANET · AODV · DSR · Energy efficient

1 Introduction

In the evolution of the wireless systems and networks, the initial realization is being achieved in the domain of wireless communication for both cellular and wireless local area network, where the only in the very last link between the base station and the wireless end system the last link is established wirelessly that offers various advantages [1]. In order to extend its capabilities and achieve a higher level of scalability, the multihop wireless networks where one or more than one intermediate node along the path that receives and forwards packets through a wireless link are being realized [2]. There are many advantages of adopting the multihop wireless networks which include: (1) There are several advantages of multihop wireless networks as compared to the single wireless links, (2) multihop wireless extends the coverage of a network, (3) improves connectivity, (4) transmission over multiple short links requires less transmission power/energy as compared to long links, (5) enables higher data rates, so there will be higher throughput and will have higher and efficient use of the wireless medium, (6) it avoids wide deployment of cables, and it

S. S. Joshi (✉) · S. R. Biradar
SDM College of Engineering and Technology, Dharwad, Karnataka, India
e-mail: sachinjoshi055@gmail.com

H. K. D. Sarma et al. (eds.), *Trends in Communication, Cloud, and Big Data*, Lecture Notes in Networks and Systems 99, https://doi.org/10.1007/978-981-15-1624-5_2

can be deployed in cost-effective ways, (7) particularly in case of denser network, several possible paths are available for the data transfer from source to destination, thus the possibility of multipath increases the robustness of the network, and thus with these capabilities' scalability, efficient and effective system is achieved [3].

The conventional routing protocols developed for fixed, cellular networks or of Internet is not suitable as well as optimal for the multihop networks, therefore in order to achieve the optimal network performance especially in case of MANET where the presence of mobility of nodes along with noisy links is a continuous evolving open research issue by optimizing various factors like bandwidth, cost, energy, etc. [4]. As well as there are many conceptualizations of the new applications which require customized routing protocol to meet their application objectives so require more efficient and robust as well as reliable customized broadcasting and multicasting routing protocols, especially for mobile wireless networks or in MANET [5, 6].

There exists a real experimental implementation of popular routing protocols including (1) Adhoc On Demand Distance Vector (AODV) Routing Protocol, (2) OLSR: Optimized Link State Routing protocol, (3) DSR: Dynamic Source Routing, etc. The MANET working committee of IETF continuously works toward achieving the goal of (1) constructing a stabilized route in the presence of dynamic topology and (2) minimization of control overhead in wireless networks without affecting the quality of potential routing behavior in wireless networks. Section 2 conducts an in-depth critical review of the literature related to the routing protocol evolutions for the MANET.

2 Review of Literature

This section of the paper describes the latest work in the field of reactive protocol especially in AODV and DSR in the last four years' time. Therefore, the initial part of the literature shows the related work to AODV, and the later part to DSR with the initial few archival works in wireless networks.

In the work of Woungang et al. [7], emphasis on the peer-to-peer networks to exploit the advantages of wireless networks for high-dimensional data content deliveries like multimedia contents to handle the problem of identification and retrieval of the data in the context of lack of routing information and lookup failures. The robust mobile chord (MR-Chord) is designed to maintain the using hash-based approach to control the effect of node mobility to get improvised lookup service rate along with improved consistency, delay, hop count and network load. The route discovery process minimization is the major challenge in the adhoc networks because the bandwidth limitations pose a typical bottleneck constraint on the QoS. This challenge requires a robust method of computation of correct expiry often route. Rios [8] in their work proposes a variable timeout that applies to any adhoc routing as well in AODV, and they claim improvised routing overhead by 8.5% minimization and approximately more than 20% lower end-to-end delay. Handling the topology storage overhead due to the dynamic aspect of the MANET, geographical routing protocol (GRP) is designed. Torrieri et al. [9] have done a comparative analysis of two different GRP along with AODV. Using the dual method of closed-form analysis for the path, delay, number hops, etc. is demonstrated with new metrics of interference correlation, fading and guard zones which are considered and shown improvements into baseline AODV. Further in the work of Alves Junior and Wille [10], an application of MANET special case VANET utilizes a customized AODV as combining features of live backbone with mobility for better connectivity as P-AODV. Selection of the appropriate routing protocol in many of the MANET-based applications requires a comparative analysis of the existing protocols. Ferronato and Trentin [11] and Biradar et al. [12] have done comparative analysis among OSLR, AODV, ZRP and AODV and AOMDV, respectively. An extensive survey is conducted by Joshi and Biradar [13], to review toward minimizing the routing overheads with consideration of RSS in the objective function formulations. The study of the frequent change's topology effect on the routing behaviors and performance remains as an open research issue as new approaches of computational intelligence like swarm optimization and machine learning impose new dimension to the problems, one of such work by Jinil Persis and Paul Robert [14], where they have modified core AODV using swarm intelligence of ANT, BEE Colony, etc., and claims improvised network performance. Such many joint studies are being started for bandwidth, energy and other parameters optimization along with the QoS of network performance, one of such work is being studied for routing overhead with energy by Joshi and Biradar [15]. Next paragraphs emphasize more on the flavor of DSR another popular reactive routing protocol for the MANET.

The poor anti-fading and slow convergence are being improvised in DSR, Meng and Song [16] in their work propose a search algorithm using Grover's theory for the optimal route section and simulation shows that network lifetime is extended as compared to conventional DSR. MANET is useful in quick deployment in case of disasters but very challenging for optimal route discovery, in the work of Karia and Godbole [17] ant colony optimization is used and proposes net protocol AntNet and compared its performance with DSR and AOMDV for performance metrics like delay, throughput, packet delivery ratio and route cost. Wang et al. [18] have exploited the use of opportunistic approaches to propose a lightweight source routing to minimize the overhead. Shivashankar et al. [19, 20] contributed to design energy-efficient routing protocol. There has been hardly more contribution to this approach, so the paper presents an energy consideration-based routing protocol design.

3 EERP System Model

The node deployment of the MANET is done randomly in the specified simulation area of $A \times A$, with N number of nodes adopting a default mobility model. The MANET is represented by GG (V, E), a graph of Graphs with V as a vertex (node) and E as an edge (radio link). The process of routing for both AODV and DSR adopted is for without energy model and with energy model. Section 3.1 describes the EERP energy model with all the constituent details of the node configuration.

$$\text{Manet} : \leftarrow \text{Graph of Graphs}\ (V,\ E)$$

The descriptive energy model for EERP indicates the field to be customized in the routing process consideration.

3.1 EERP Energy Model

The core baseline protocol of both AODV and DSR mainly focuses on the route discovery process along with route maintained as an effort of fault tolerance, whereas the consideration of energy while these processes are not considered. In the proposed EERP model, the node configuration of a wireless node/MANET node is enabled for the energy by bringing change in the base model, where the initial energy is defined as infinity so that no log for the energy is created as it is not considered as a constraint. Figures 1 and 2 show both the node configuration format without and with energy consideration, respectively.

The attributes of the energy model in a node show the energy level in a mobile host at the beginning of the communication cycle is usually termed as initial energy. Another important attribute related to energy includes (1) energy usage for each packet for both transmission and receiving terms as P.Tx and P.Rx, respectively. The initial attributes of energy are considered in Joule, and the remaining other energy parameters are considered in Watts.

The display model of the protocol implemented in NS2—a discrete event simulator for energy model is shown in Fig. 3 for the different scenarios.

The typical AODV and DSR follow the IETF standard for executing route discovery operation and route maintenance process, and the trace route file of the process is

Address type: Hierarchical	Adhoc Routing: AODV	
mac Type: MAC:802.11	ifq Type: Queue/ifqLen50	
Antenna type: Omni Antenna	Propagation type: TwoRayGround	
Physical layer type: Phy/WirelessPhy	TopologyInstance: topo	
Channel: WirelessChannel	agent Trace: ON	
Router Trace: ON	mac Trace: OFF	movement Trace: OFF

Fig. 1 Node configuration without energy model

Address type: Hierarchical	Adhoc Routing: AODV	
mac Type: MAC:802.11	ifq Type: Queue/ifqLen50	
Antenna type: Omni Antenna	Propagation type: TwoRayGround	
Physical layer type: Phy/WirelessPhy	topologyInstance: topo	
Channel: WirelessChannel	agent Trace: ON	
Router Trace: ON	macTrace: OFF	MovementTrace: OFF
Initial Energy: 90	Transmission power :0.5	
Transmission power: 0.3	idle power: 0.05	Sleep power: 0.03

Fig. 2 Node configuration with energy model

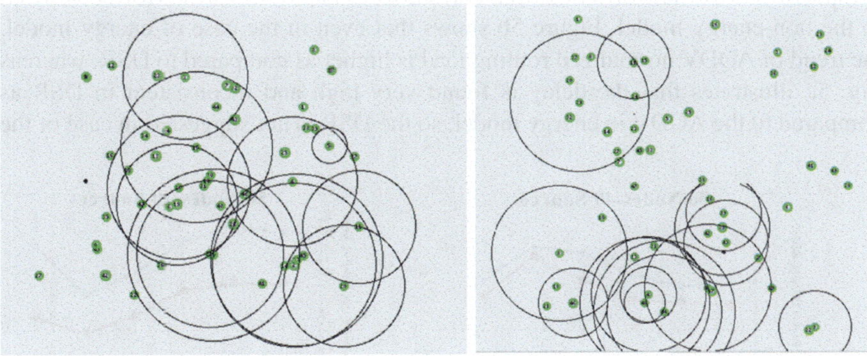

EN-AODV-50nodes--40sources-0_Pause_Time

EN-DSR-AODV-50nodes--40sources-1000_Pause_Time

Fig. 3 Network communication scenario in NS2 for both AODV and DSR

retained as a log in the NS2 environment for analysis of various network performance parameters for AODV and DSR with both energy and without energy models.

4 Result and Analysis

Figure 4 illustrates multiple graphs where Fig. 4a shows the performance graph of packet delivery ratio (PDR), Fig. 4b displays the performance graph of normalized routing load and Fig. 4c illustrates the performance graph of delay with respect to pause time.

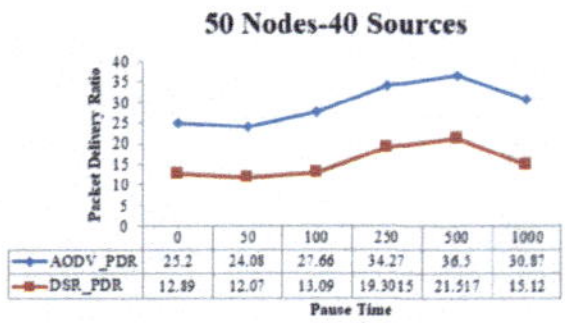

(a) Packet Delivery Ratio Vs. Pause Time

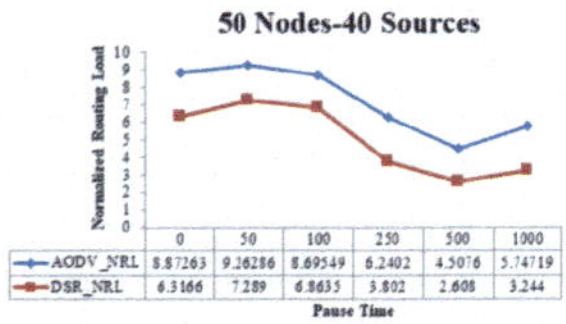

(b) Normalized Routing Load Vs. Pause Time

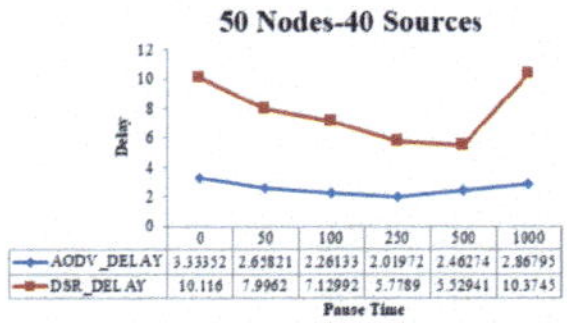

(c) Delay Vs. Pause Time

Fig. 4 AODV and DSR without energy

It is found that the PDR of AODV is higher as compared to DSR, the normalized routing load which is a measure of bandwidth or traffic is found higher for the AODV and also the delay is more and inconsistent for AODV as compared to DSR. Therefore, AODV outperforms comparatively to the DSR without energy model.

Figure 5 illustrates four different performance graphs; Fig. 5a illustrates the PDR of the AODV with energy model having a higher value at different pause time as compared to DSR. It is also observed that the DSR improvises its PDR as compared to the non-energy model. Figure 5b shows that even in the case of energy model, the trend of AODV normalized routing load is higher as compared to DSR, whereas Fig. 5c illustrates that the delay is found very high and inconsistent in DSR as compared to the AODV in energy model, so the DSR is not suggested in case of the

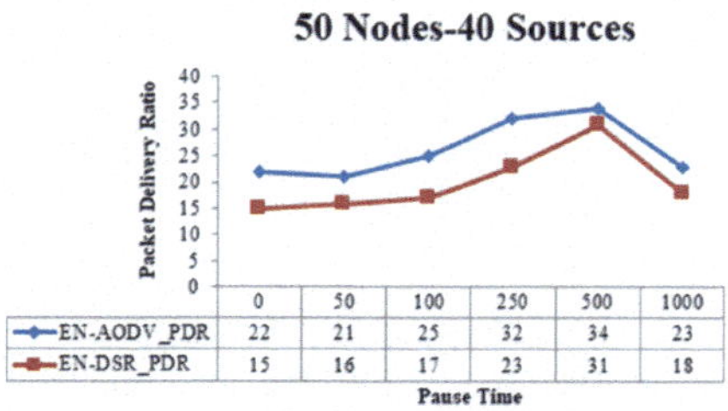

(a) Packet Delivery Ratio Vs. Pause Time

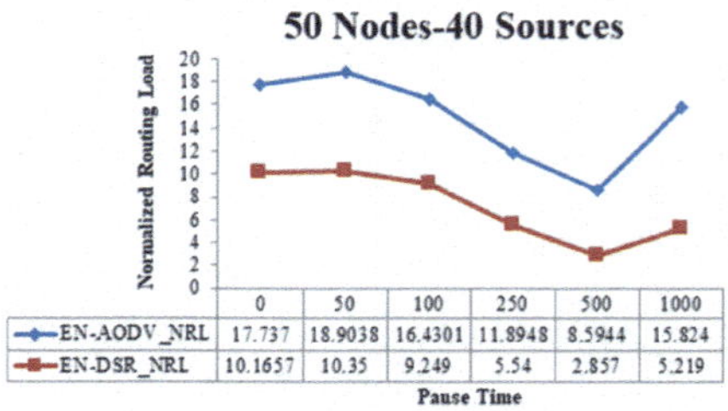

(b) Normalized Routing Load Vs. Pause Time

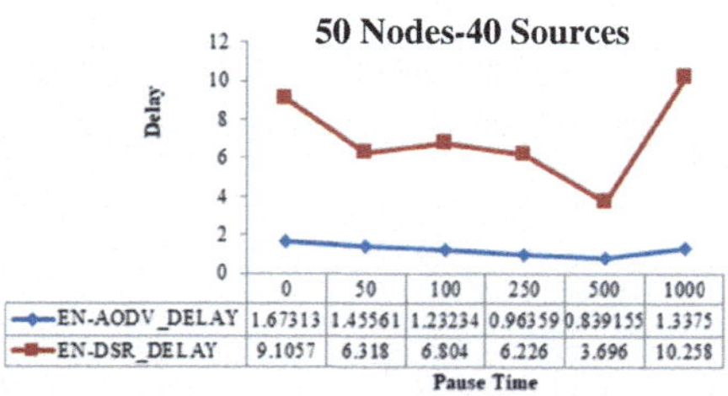

(c) Delay Vs. Pause Time

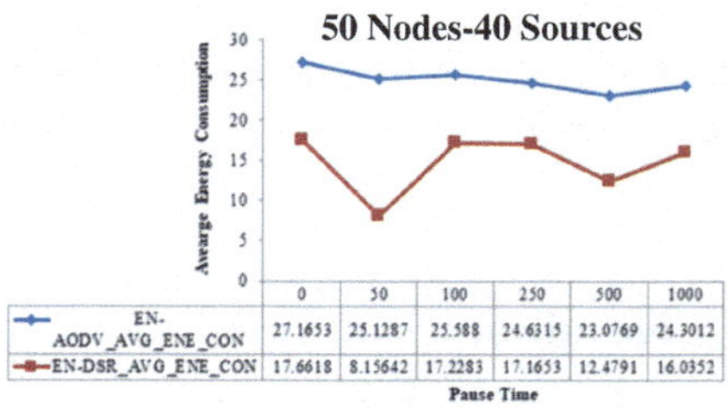

(d) Average Energy Consumption Vs. Pause Time

Fig. 5 AODV and DSR with energy model

reliability-based application of MANET. The most important observation found is that AODV takes more energy as compared to DSR.

5 Conclusion and Future Research Direction

The MANET is very promising adhoc technology-based applications to provide a platform for both independent applications as well as joining subnet in the edges of the internet of things or on the internet. The efforts of the routing protocol development are challenging because of its dynamic topology. The conventional AODV and DSR are a reactive protocol which assumes energy saving, but its code deployment focuses mainly on the route discovery process. This paper proposes a modified AODV and DSR, namely EERP: Energy-Efficient Routing Protocol Comparative Study of DSR and AODV for the mobile adhoc network. The modified AODV and DSR are simulated on discrete event simulation on NS2 with energy parameters consideration, and AODV performs better comparatively as well as DSR also improvises its performance as compared to the model without energy. In the future more, specific topology along with another network constraint consideration, the routing protocol could be modified to have its synchronous advantages in the real-time scenarios.

References

1. Hara T, Zadorozhny VI, Buchmann E (eds) (2010) Wireless sensor network technologies for the information explosion era, vol 278. Springer
2. Bernas M, Toyonaga S, Kominami D, Sugano M, Mesin L, Hasani B (2018) Advances in wireless sensor network. In: Advances in wireless sensor network, 361 pp
3. Rani S, Ahmed SH (2015) Multi-hop routing in wireless sensor networks: an overview, taxonomy, and research challenges. Springer
4. Kumar M, Sharma C, Dhiman A, Rangra AK (2018) Performance variation of routing protocols with mobility and scalability in MANET. In: Next-generation networks. Springer, Singapore, pp 9–21
5. Lee S-J, Su W, Gerla M (2002) On-demand multicast routing protocol in multihop wireless mobile networks. Mob Netw Appl 7(6):441–453
6. Kihl M, Sichitiu M, Ekeroth T, Rozenberg M (2007) Reliable geographical multicast routing in vehicular adhoc networks. In: Wired/wireless internet communications. Springer, Berlin, Heidelberg, pp 315–325
7. Woungang I, Tseng F, Lin Y, Chou L, Chao H, Obaidat MS (2015) MR-Chord: improved chord lookup performance in structured mobile P2P networks. IEEE Syst J 9(3):743–751
8. Rios M (2015) Variable route expiration time based on a fixed probability of failure for adhoc networks routing applications. IEEE Lat Am Trans 13(1):383–389
9. Torrieri D, Talarico S, Valenti MC (2015) Performance comparisons of geographic routing protocols in mobile adhoc networks. IEEE Trans Commun 63(11):4276–4286
10. Alves Junior J, Wille ECG (2015) Increasing connectivity in VANETs using public transport backbones. IEEE Lat Am Trans 13(10):3421–3431
11. Ferronato JJ, Trentin MAS (2017) Analysis of routing protocols OLSR, AODV and ZRP in real urban vehicular scenario with density variation. IEEE Lat Am Trans 15(9):1727–1734

12. Biradar SR, Singh SK, Sarkar SK (2011) Performance comparison of AODV and AOMDV routing protocols. In: Proceedings of the international conference on wireless networks (ICWN) 2011. The Steering Committee of the World Congress in Computer Science, Computer Engineering and Applied Computing (WorldComp), p 1
13. Joshi SS, Biradar SR (2014) Reviewing contribution for mitigating routing overhead and adopting signal strength in MANET. Int J Comput Appl 103(10)
14. Jinil Persis D, Paul Robert T (2017) Review of adhoc on-demand distance vector protocol and its swarm intelligent variants for mobile adhoc network. IET Netw 6(5):87–93
15. Joshi SS, Biradar SR (2016) Communication framework for jointly addressing issues of routing overhead and energy drainage in MANET. Procedia Comput Sci 1(89):57–63
16. Meng L, Song W (2013) Routing protocol based on Grover's searching algorithm for mobile adhoc networks. China Commun 10(3):145–156
17. Karia DC, Godbole VV (2013) New approach for routing in mobile adhoc networks based on ant colony optimisation with global positioning system. IET Netw 2(3):171–180
18. Wang Z, Chen Y, Li C (2014) PSR: a lightweight proactive source routing protocol for mobile adhoc networks. IEEE Trans Veh Technol 63(2):859–868
19. Shivashankar S, Suresh HN, Varaprasad G, Jayanthi G (2014) Designing energy routing protocol with power consumption optimization in MANET. IEEE Trans Emerg Top Comput 2(2):192–197
20. Shivashankar S, Varaprasad G, Narayanagowda SH (2014) Implementing a new power aware routing algorithm based on existing dynamic source routing protocol for mobile adhoc networks. IET Netw 3(2):137–142

Mathematical Modeling of Mobility Models used in IP-Based Wireless Mobile Networks

Kushal Pokhrel, Nitul Dutta, M. K. Ghose and Hiren Kumar Deva Sarma

Abstract Wireless mobile networks (WMNs) should be tested under various realistic conditions including transmission range, type and pattern of data traffic and movement pattern of the users. This paper is a review of mathematical modeling of mobility models that cater to the movement pattern of the users in IP-based WMNs. We describe five mobility models where movement pattern of the mobile user is estimated by some probability distribution function. The aim of this paper is to present some of mobility models based on classical theories in order to support researchers with options to choose when their protocol or architecture needs to be studied under various mobility patterns.

Keywords Mobility models · Mobility management protocols · IP-based wireless mobile networks

1 Introduction

The mobility model is one of the key determinants in the accurate approximation of the movement of a mobile node (MN) in IP-based wireless mobile networks. The underlying mobility model essentially mimics the movement behaviors of actual users. Some mobility models are simple but do not represent the approximately

K. Pokhrel (✉)
Department of CSE, SMIT, Majitar, Sikkim, India
e-mail: kushal.p@smit.smu.edu.in

N. Dutta
Faculty of Technical Sciences, Jan Wyzykowski University, Ul. Skalnikow 6 B, 59-101 Polkowice, Poland
e-mail: n.dutta@ujw.pl

M. K. Ghose
CSE Department, Om Dayal Group of Institutions, Howrah, West Bengal 711316, India
e-mail: mkghose2000@yahoo.com

H. K. D. Sarma
Department of IT, SMIT, Majitar, Sikkim, India
e-mail: hiren.s@smit.smu.edu.in

H. K. D. Sarma et al. (eds.), *Trends in Communication, Cloud, and Big Data*, Lecture Notes in Networks and Systems 99, https://doi.org/10.1007/978-981-15-1624-5_3

real user movement patterns, while others being complex provide more realistic movement patterns. Therefore, mobility modeling is a critical tool for supporting realistic and accurate protocol simulations of movement of users in IP-based wireless mobile network (WMN).

The movement of MNs in a WMN is usually random in nature and the direction in which the nodes move that are quite uncertain. Many literatures have been reported over the years, for different mobility models applicable for different types of wireless networks [1–3]. Some mobility models are purely synthetic models, while some are trace based; at the same time, some are network based [4]. The implementation of different mobility management protocols is very difficult as it is tricky merely to be able to distinguish the applicability of these different models under different network architectures. Our work deals with some existing mobility models with their importance to evaluate protocols or architecture. The mobility models are designed to describe the movement pattern of MNs, and how their location, speed and acceleration change over time. If a mobility model reasonably fails to emulate the movement patterns of MNs in a real-time environment, then the conclusions drawn from the simulation studies may be misleading. Therefore, the choice of mobility management protocol is quite critical in simulation research of WMNs. Furthermore, it is also very vital to define a metric for the so-called degree of mobility of the MN as there are broad ranges of mobility models used in the literature and their various parameters and such a definition is not trivial. In this paper, we have tried to address and realize a mathematical model of some of the mobility parameters of major interests like speed behavior, direction change behavior, transition length, transition time and pause time.

The rest of the paper is organized as follows:

In Sect. 2, we discuss mobility models to be used in IP-based WMNs, and then from Sects. 2.1–2.5, we shall discuss five such mobility models. The work discussions are provided in Sect. 3 wherein we have discussed all the mathematical models. In Sect. 4, we present the conclusions citing the possible future extensions of the work.

2 Mobility Models

In this section, we shall provide approximate mathematical models of few mobility models based on classical theories. We are discussing four mobility models with their characteristics and also present how they can be used in simulation or mathematical modeling of the mobile network-based protocols.

2.1 Fluid-Flow Mobility Model

Fluid-flow mobility model [5] describes macroscopic movements instead of individual or microscopic movements. The behavior of the generated traffic is similar to a fluid flowing through a pipe. As a result, the fluid-flow mobility model best represents traffic on highways and other similar situations with a constant flow of MNs; in other words, the model is unable to accurately represent the movements of individual MNs. The fluid-flow model is more suitable for users with high mobility, infrequent speed and direction changes [6]. The average location update rate, $E\left[R_{A(X)}\right]$, is given by (1).

$$E\left[R_{A(X)}\right] = \frac{\bar{v}L(K)}{\pi S(K)} \tag{1}$$

where

v	be the average speed in km/h
$S(K)$	is the area of region $A(K)$
$L(K)$	is the perimeter of the region $A(K)$
$A(K)$	is the region which is uniformly distributed in the range $(0, 2\pi)$.

2.2 *Random Waypoint Model*

In this mobility model [7], i.e., random waypoint (RWP) model, a node movement is studied till its destination based on its transition length and time, spatial distribution, direction angle and cell change rate in a system area. The RWP model is described as a discrete-time stochastic process of movements in one-dimensional (i.e., line segment) and two-dimensional (i.e., rectangle or circle).

The probability density function of the transition length L_R of a MN moving according to the RWP model on a disk of radius a is given by

$$f_{l_R}(l_R) = \frac{4}{\pi a^2}\left[\operatorname{Cos}^{-1}\left(\frac{1}{2a}\right) - \frac{1}{2a}\sqrt{1-\left(\frac{l}{2a}\right)^2}\right] \tag{2}$$

Now assuming the pause time T_p at the waypoints is exponentially distributed, with the arrival rate of λ_R the pdf of pause time T_p can be modeled as

$$f_{T_p}(t_p) = \lambda_R e^{-\lambda t_p} \quad \text{if } t_p \geq 0 \text{ and } 0 \text{ otherwise} \tag{3}$$

In a RWP model, the MN is assumed to be moving with a constant speed V from one waypoint to the other. Therefore, the transition time can be modeled as

$$T_R = \frac{L}{V} \tag{4}$$

The total process time may be modeled as

$$T_{RWP} = T_R + T_p \tag{5}$$

The direction that the MN adopts from a given waypoint within the circular disk of radius a is the function of the angle $\theta_R \epsilon (0 < \theta_R < 2\pi)$ which is assumed to be uniformly distributed. The pdf of the direction from the waypoints can be modeled as in (8)

$$f_{\theta_R}(\theta_R) = \frac{L}{2\pi} \tag{6}$$

for $(0 \leq \theta_R \leq 2\pi)$ and 0 otherwise.

In RWP model, the change in direction of a MN depends upon the speed of the MN and the shape of the system area. For a given area, the frequency of direction change by the mobile node is directly proportional to the speed of the MN. Thus, the time between two successive direction changes may also be determined by (7). Additionally, we can write

$$E\{T\} = E\{T_{\mathrm{RWP}}\} = E\{T_R\} + E\{T_p\} \tag{7}$$

where $E\{T\}$ represents the average time between each direction change. Alternately, the direction change frequency is given by $\frac{1}{E\{T\}}$unit/s.

Our first aim is to analyze how many cell boundaries a node crosses on average during one movement transition. The corresponding random variable is denoted by C. By denoting c_i as the outcome of this random variable in transition i, we can write

$$E\{C\} = \lim_{m \to \infty} \frac{1}{m} \sum_{i=1}^{m} c_i \tag{8}$$

The expected number of cell changes per unit time is the sum of all cell changes for a given time. We write:

$$E\{C_t\} = \lim_{m \to \infty} \frac{\sum_i^m C_i}{\sum_i^m T_i} = \frac{E\{C\}}{E\{T\}}. \tag{9}$$

2.3 The Gauss–Markov Mobility Model

In this mobility model, the instantaneous value of speed (s) and direction (d) of the MN at the nth instant is calculated based upon the value of speed and direction at the $(n-1)$th instant and a random variable using the following equations:

$$s_n = \alpha s_{n-1} + (1-\alpha)\bar{s} + \sqrt{(1-\alpha^2)} s_{x_{n-1}} \tag{10}$$

$$d_n = \alpha d_{n-1} + (1-\alpha)\bar{d} + \sqrt{(1-\alpha^2)d_{x_{n-1}}} \tag{11}$$

where s_n and d_n are the new speed and direction of the mobile user at time interval n. α, $(0 \le \alpha \le 1)$, is the tuning parameter used to vary the randomness; s and d representing the mean value of speed and direction and constants. $s_{x_{n-1}}$ and $d_{x_{n-1}}$ are random variables from a Gaussian distribution. A total random values (or Brownian motion) are obtained by setting $\alpha = 0$, and linear motion is obtained by setting $\alpha = 1$ as reported in [8]. Intermediate levels of randomness are obtained by varying the value of α between 0 and 1. The mean speed of the MN in Gauss–Markov mobility model is given by

$$s_{\text{in}} = \sum_n \frac{s_n}{n} \tag{12}$$

The average distance traveled by MN in nth transition from origin is given by

$$d_{\text{in}} = \sum_n \frac{\sqrt{(y_n - y_i)^2 + (x_n - x_i)^2}}{n} \tag{13}$$

Therefore, the mean cell change rate for Gauss–Markov mobility model may be expressed as the sum of all arrival rates into different cells [9] expressed as follows:

$$\text{CCR}_{\text{GMM}} = \sum_{j=1}^{n} \frac{s_{\text{in}}}{d_{\text{in}}} \tag{14}$$

2.4 *The Brownian Motion Mobility Model*

The Brownian motion mobility model is based on Einstein's classical theory of Brownian motion explained in [10]. According to this theory, the probability of n discrete observations in duration Δt_{B} can be given expressed as:

$$P(X(t_{\text{B}}) = n) = P_n(t_{\text{B}}) = \left(\frac{e^{-\lambda_{\text{B}}}\lambda_{\text{B}}^n}{n!}\right) \tag{15}$$

where λ_{B} is the average arrival rate of MN, and $X(t_{\text{b}})$ is a stochastic process which expresses the number of observations in time t_{b}.

Now let $X_i(t_{\text{B}})$ for $i = 1, 2, 3, \ldots N_{\text{B}}$, be independent Poisson processes with average arrival rate λ_{B} as described above. Let $Y_N(t_{\text{B}})$ be another stochastic process defined by

$$Y_{N_{\mathrm{B}}}(t_{\mathrm{B}}) = \sum_{i=1}^{N_B} \frac{X_i(t_{\mathrm{B}}) - \lambda_{\mathrm{B}} \Delta t_{\mathrm{B}}}{\sqrt{\lambda_{\mathrm{B}} N_{\mathrm{B}}}} \tag{16}$$

The total transition time may be modeled as follows:

$$T_{\mathrm{B}} = n\Delta t_{\mathrm{B}} + \sum_{n=1}^{N} \left\{ W\left(t_{\mathrm{B}_n}\right) - W\left(t_{\mathrm{B}_{n-1}}\right) \right\}; \quad t_{\mathrm{B}_N} \in \tau \tag{17}$$

Considering a circular area of radius a and assuming the mean transition length as $\overline{l_{\mathrm{B}}} = a$ using the same argument as before, the transition length (total displacement of MN) between the points $W(0) = 0$ (origin) to $W(C)$, within the circular area may be modeled as.

$$L_{\mathrm{B}} = a\sqrt{x^2 + y^2}; \quad 0 < L_{\mathrm{B}} < 2a \tag{18}$$

where (x, y) are the coordinate axes of the point $aW(t_{\mathrm{B}})$; at $t_{\mathrm{B}} = C; \quad C \in \tau.$

The speed of MN during transition may be assumed to follow a normal distribution which can be modeled as follows:

$$f_{V_{\mathrm{B}}}(v_{\mathrm{B}}) = \int_{0}^{\frac{L_{\mathrm{B}}}{T_{\mathrm{B}}}} \frac{1}{\sqrt{2\pi\sigma^2}} \mathrm{e}^{-\frac{(v_{\mathrm{B}} T_{\mathrm{B}} - L_{\mathrm{B}})^2}{2T_{\mathrm{B}}^2 \sigma^2}} \mathrm{d}v_{\mathrm{B}} \tag{19}$$

where the mean has been assumed to be $\mu = \frac{L_{\mathrm{B}}}{T_{\mathrm{B}}}$ and σ is the standard deviation. The cell change rate handover rate in a circular area of radius a can be thought as the fraction of the total mean epoch displacement and the total transition time expressed as:

$$\mathrm{CCR}_{\mathrm{B}} = \frac{\int_0^a L_{\mathrm{B}}(t)\mathrm{d}t}{2aT_{\mathrm{B}}} \tag{20}$$

2.5 Markovian Waypoint Mobility Model (MWP)

In this mobility model, the MN is thought to move along a straight line segment from the current waypoint (x_i, y_i) to the next (x_{i+1}, y_{i+1}) [11]. The waypoint coordinates $(x_0, y_0); (x_0, y_1)\ldots$ constitute a Markov chain inside a circular area with radius a satisfying the equation $x^2 + y^2 = a^2$ with transition probability density matrix $p_{x,y}$. The waypoints given by coordinates (x_{i+1}, y_{i+1}) can be found out using the probability density matrix $p_{x,y}$ and the previous waypoints coordinates (x_i, y_i), the conditional pdf of the same may be modeled as:

$$P(X_{i+1}, Y_{i+1}) = \int_{0}^{\sqrt{a^2-y^2}} \mathrm{d}x \int_{0}^{\sqrt{a^2-x^2}} p_{x,y} P(X_i, Y_i)\mathrm{d}y \quad (21)$$

where the transition probability matrix $p_{x,y}$ is having the following properties:

The transition length L_M may be modeled as

$$L_\mathrm{M} = \sqrt{(x_{i+1} - x_i)^2 + (y_{i+1} - y_i)^2}; \quad 0 < L_\mathrm{M} < 2a \quad (22)$$

Assuming the number of instants N_M, the MN takes pause during the transition to be a discrete Poisson distribution with mean λ_{N_M}, the pmf of the function for the number of pauses taken can be given as follows:

$$f_{N_\mathrm{M}}\left(n, \lambda_{N_\mathrm{M}}\right) = \sum_{n=0}^{\infty} \left(\frac{\mathrm{e}^{-\lambda_N} \lambda_{N_\mathrm{M}}^{\mathrm{n}}}{n!} \right); \quad n = 1, 2, 3, 4 \ldots \quad (23)$$

Now assuming the pause times T_t during transitions are exponentially distributed, with the arrival rate of λ_P the pdf of pause time T_t can be modeled as follows:

$$f_{T_\mathrm{t}}(t_\mathrm{t}) = \lambda_P \mathrm{e}^{-\lambda t_t} \quad \text{if } t_t \geq 0 \text{ and } 0 \text{ otherwise} \quad (24)$$

The direction that the MN adopts from a given waypoint within the circular disk of radius a is the function of the angle $\theta_\mathrm{M} \in (0 < \theta_\mathrm{M} < 2\pi)$ which is assumed to be uniformly distributed. The pdf of the direction from the waypoints can be modeled as follows:

$$f_{\theta_\mathrm{M}}(\theta_\mathrm{M}) = \frac{L_\mathrm{M}}{2\pi} \text{ for } (0 \leq \theta_\mathrm{M} \leq 2\pi) \text{ and } 0 \text{ otherwise} \quad (25)$$

The speed of MN during transition may be assumed to follow a normal distribution which can be modeled as follows:

$$f_{V_\mathrm{M}}(v_\mathrm{M}) = \int_{0}^{\frac{L_\mathrm{M}}{T_\mathrm{t} N_\mathrm{M}}} \frac{1}{\sqrt{2\pi\sigma^2}} \mathrm{e}^{-\frac{L_\mathrm{M}}{4T_\mathrm{t}\sigma^2 N_\mathrm{M}}} \mathrm{d}v_M \quad (26)$$

where the mean has been assumed to be $\mu = \frac{L_\mathrm{M}}{2T_\mathrm{t} N_\mathrm{M}}$ and σ is the standard deviation.

The total process time in a MWP model may thus be modeled as follows:

$$T_\mathrm{MWP} = N_\mathrm{M} T_\mathrm{t} + T_\mathrm{p} \quad (27)$$

The cell change rate for Markovian mobility model may be expressed in the same way as for Brownian mobility model as follows:

$$\mathrm{CCR}_{\mathrm{MWP}} = \frac{\sum_i L_{\mathrm{M}_i}}{2aT_{\mathrm{MWP}}} \tag{28}$$

3 Discussions

It is very much evident from earlier deliberations that mobility management is a key metric to evaluate the performance of WMNs. As it is vital that the mobility model used should be able to approximately emulate the movement pattern of MNs in a real IP-based WMN. In this regard, we have provided the much needed mathematical analysis for five different mobility models in a single platform, namely fluid flow, random waypoint, Gauss–Markov, Markovian and Brownian. We have also derived the equations for the "degree of mobility" alias the cell change rate for all the above mobility models in the expressions cited in (1), (9), (14), (20) and (28).

4 Conclusion

In this paper, five mobility models and their characteristics are discussed to enable researchers with the idea to select a suitable mobility model to test the performance of the protocol or architecture designed for mobile IP-based network. Though we target the IP-based mobile network here, these models are equally important for other wireless networks where users are mobile.

Fluid-flow mobility model rather describes macroscopic movements instead of microscopic movements. The behavior of the generated traffic is similar to a fluid flowing through a pipe, and hence, this mobility model may be applicable to estimate the movement patterns of MNs in a highway scenario. The random waypoint mobility model is flexible, and it appears to create realistic mobility patterns for the way people might move in. It is suitable to model a scenario of a conference room or small office premises where users are pedestrian. The Gauss–Markov mobility model provides movement patterns that one might expect in the real world, provided appropriate parameters are chosen. The method can be used to force users to go away from the boundary of the system area during simulation which may avoid undesired edge effects. The Brownian mobility model follows a very random movement of the MNs specially envisaged in terms of high gathering of people. The Markovian waypoint mobility model caters to the movement of a MN in straight lines in a confined cell. The essence of this mobility model lies in the fact that the subsequent movement patterns are memoryless in nature.

Therefore, we conclude by stating that the expressions derived in our work give a comprehensive mathematical analysis of the critical mobility models that may be implemented various WMNs. The future extensions of the work may be to implement the expressions derived to generate mobility pattern results by simulation.

References

1. Einstein A (1956) Investigations on the theory of the Brownian movement. Dover Publications
2. Nain P, Towsley D, Liu B, Liu Z (2005) Properties of random direction models. In: Proceedings of INFOCOM'05
3. Johnson DB, Maltz DA (1996) Dynamic source routing in ad hoc wireless network. In: Imielinski T, Korth H (eds) Mobile computing, Chapter 5. Kluwer Academics Publisher, pp 153–181
4. Musolesi M, Mascolo C (2008) Mobility models for systems evaluation a survey. http://www.cl.cam.ac.uk/~cm542/papers/minema08.pdf. Accessed on 29th Sept 2016
5. Dutta N, Misra IS (2012) Fluid flow and random walk mobility model for wireless mobile network research: a review. Adv Mater Res 403–408:4486–4494
6. Lam D et al (1997) Teletraffic modeling for personal communications services. IEEE Commun Mag
7. Bettstetter C (2004) Stochastic properties of the random waypoint mobility model. In: Wireless networks, vol 10. Kluwer Academic Publishers, The Netherlands, pp 555–567
8. Li J, Jannotti J, Couto D, Karger D, Morris R (2000) A scalable location service for geographic ad hoc routing. In: Proceedings of ACM international conference on mobile computing and networking (MobiCom), Boston, MA
9. Liang B, Haas Z (1999) Predictive distance-based mobility management for PCS networks. In: Proceedings of the joint conference of the IEEE computer and communications societies (INFOCOM)
10. Camp T, Boleng J, Davies V (2002) A survey of mobility models for ad hoc network research. Wirel Commun Mobile Comput 2(5):483–502
11. Tolety V (1999) Load reduction in ad hoc networks using mobile servers

A Game Theory-Based Framework for Reliable and Energy-Efficient Data Delivery in Cognitive Radio Wireless Sensor Network

Prativa Rai, M. K. Ghose and Hiren Kumar Deva Sarma

Abstract Depending on application type, cognitive radio wireless sensor network (CRWSN) needs energy-efficient data delivery mechanism that can ensure certain level of reliability. Game theory may be exploited for designing network protocols with different objectives. In this work, a game theory-based framework has been developed that can be used for energy-efficient and reliable data delivery in CRWSN. For this purpose, game theory-based channel allocation and routing protocols for CRWSN shall be developed, in the future. Implementation plan of the proposed framework has been outlined.

Keywords Wireless sensor network · Cognitive radio · Cognitive radio wireless sensor network · Spectrum holes · Channel assignment · Routing

1 Introduction

All the wireless applications use radio spectrum for communication which is a natural resource and has become very scarce due to the growth of wireless applications. To use the radio spectrum, the government auctions the frequency bands and grants licenses in a long-term basis. The company which gets the licensed over the frequency bands is called licensed users or primary users (PU). Due to the growth of wireless applications, the request for spectrum allocation has increased exponentially. However, the utilization of the licensed spectrum is quite low and uneven, leading to a waste of valuable frequency resources, as per the studies of Federal Communication

P. Rai (✉)
Department of Computer Science & Engineering, Sikkim Manipal Institute of Technology, Sikkim Manipal University, Majitar, Sikkim 737136, India
e-mail: prativa.r@smit.smu.edu.in

M. K. Ghose
Computer Science & Engineering, OmDayal Group of Institutions, Uluberia, Howrah, West Bengal 711316, India

H. K. D. Sarma
Department of Information Technology, Sikkim Manipal Institute of Technology, Sikkim Manipal University, Majitar, Sikkim 737136, India

H. K. D. Sarma et al. (eds.), *Trends in Communication, Cloud, and Big Data*, Lecture Notes in Networks and Systems 99, https://doi.org/10.1007/978-981-15-1624-5_4

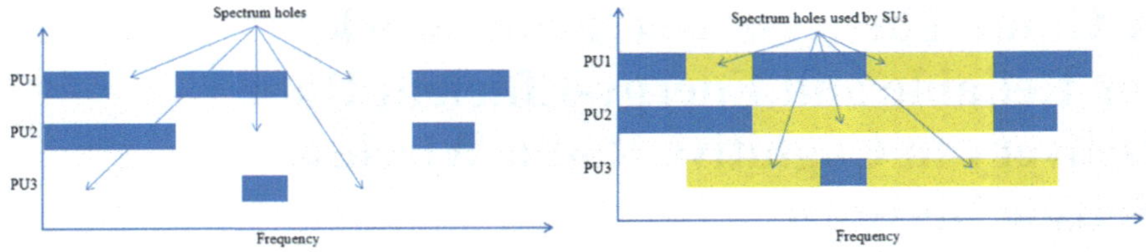

(a) Frequency usage by PU & Spectrum Holes **(b) Use of Spectrum Holes by SUs**

Fig. 1 **a** Frequency usage by PU and spectrum holes. **b** Use of spectrum holes by SUs

Commission (FCC) [1]. The frequency bands owned by PU's which are underutilized are known as spectrum holes or white spaces as shown in Fig. 1a. On the other hand, the existing unlicensed spectrum mainly the industrial–scientific–medical (ISM) band has become overcrowded with different applications such as WiMax, Bluetooth, Wi-Fi, etc. Therefore, adding more applications in this band is purposeless. One solution to improve spectrum utilization is to allow unlicensed users to access the spectrum holes opportunistically. These unlicensed users are also called secondary users (SUs) which will utilize the unoccupied frequency band of the PU's. This concept is depicted in Fig. 1b. Therefore, the concept of cognitive radio (CR) is a potential technique for efficient utilization of spectrum resource.

A CR comes across as a smart wireless communication solution that is not only sensitive of its neighboring vicinity (self-awareness), but also has the capability to adapt its vital attributes (self-configuring) according to the environment. This essence of CR enables it to attain a reliability and proficiency [2]. In the literature, CR has been defined as a radio that is capable of offering reliable, spectrally efficient communication solution that adapts its internal parameters as per the changes in its deployed environment [3]. The CR-enabled SUs are spectrally efficient users which have the capability of using licensed bands without causing any interference to PUs. The overall spectral efficiency of the network thus increases with adoption of CR.

A wireless sensor network (WSN) comprises of numerous sensor nodes which are spatially distant and at the same time, resource constrained. They are used to sense any given environment-related real-time data. Each node in WSN senses its surroundings and links the captured data to the base station (BS). The BS after processing the captured data transmits the data for further processing by the user. The WSN's that have been deployed so far use the unlicensed band (ISM band) which is already over crowed leading to unfavorable power losses to the sensors thereby making then inefficient. There has been a significant development in both hardware and software aspects to enhance the overall performance of the WSNs but still lot of challenges remain open.

Recent studies in [4–7] have introduced CR technology to be incorporated in WSN to overcome the shortcomings of the traditional WSNs. CR permits the use of unutilized licensed frequency bands which converged with WSN enables WSN to be more reliable and energy efficient [8].

Cognitive radio wireless sensor networks (CRWSNs) are special type of wireless sensor networks in which the sensors are CR-enabled, i.e. the sensor nodes are capable of self-configuring incase of any changes in the surrounding environment [8]. Each node in CRWSN not only sends or receives data but also senses the spectrum to find spectrum holes.

A cognitive radio wireless censor network (CRWSN) may be defined as a distributed network of wireless CR-enabled sensor nodes that can sense event signals and transmit the sensed data in collaboration with the neighboring CRWSN nodes, in a dynamic manner in order to suit any specific application. The intercommunication between the CR-enabled sensor nodes and the corresponding base station in a CRWSN is carried out over the vacant spectrum bands in a multi-hop manner.

The nodes of CRWSN are different from the traditional WSN nodes, basically with the presence of the RF unit. The CR node chooses the most suitable channel among the sensed idle channels and also releases the channel in the event of a PU claiming access to the channel at any point of time. CRWSN leads to increase in spectrum utilization, increase network efficiency, and extend the lifetime of WSNs thus fulfilling end-to-end goal [4]. Figure 2 presents the architecture of CRWSNs.

With the introduction of CR functionality in WSN, CRWSN operates and behaves differently from the traditional WSN. The differences are summarized based on several factors in Table 1.

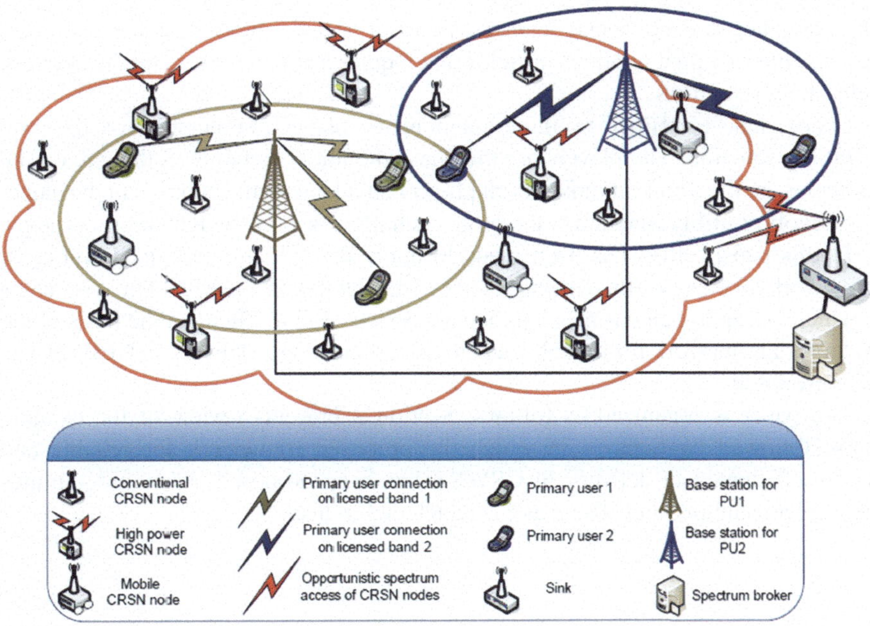

Fig. 2 CRWSN architecture Akan et al. [4]

Table 1 Difference between WSN and CRWSN

Parameters	CRWSN	WSN
Frequency band used	Licensed and ISM bands	ISM bands
Communication depends on	PUs	Not applicable
Energy consumption	Low	High
Accuracy	High	Low
End-to-end delay	Low	High
Protocol intricacy	High	Low
Network overhead	High	Low
Suitable in	Overloaded ISM band	Normal load ISM band
Utilization of spectrum holes	Yes	No
Multi-channel	Required	Possible

Game theory (GT) is a mathematical tool used for studying the conflicting behavior among different individual [9]. Numerous concepts of GT are now available in the literatures such as extensive-form games, repeated games, Shapley value, and evolutionary game theory [10]. The concept of GT has been widely used in diverse subjects like computer science, social sciences, political science, biological science, engineering, international relations or fields that require analyzing the conflicts between individuals or entities.

Every game is defined by three components: players, strategies, and the payoff/utility function. The players are the ones who are contending in the game. The strategies are the different moves each player can take with the aim to win the game. The payoff or utility function is the profit each players gains by playing its strategy.

In this paper, effort has been made to apply the GT approach in designing a framework for energy-efficient and reliable data delivery in CRWSN. Different areas in which GT approach has been applied are reviewed first. Then, on the basis of the findings, a framework for reliable and energy-efficient data delivery in CRWSN has been proposed.

The paper is organized as follows. Section 2 presents review of the existing work, followed by Sect. 3, in which the proposed framework for reliable and energy-efficient data delivery in CRWSN has been proposed. Section 4 outlines the implementation plan. The paper is concluded in Sect. 5.

2 Related Work

The concept of GT has been applied in WSN [11–17], where extensive surveys have been carried out in order to enlist the various taxonomies of existing approaches highlighting the open and future challenges. Various types of GT approaches have been demonstrated in numerous types of WSNs. The research work incorporating different GT techniques in CRWSN can be classified into following board categories:

(i) Spectrum decision and spectrum sharing leading to channel allocation,
(ii) Designing clustering algorithms,
(iii) Designing routing algorithms.

In that context, an energy-efficient spectrum decision scheme for CRWSN has been presented in [18]. The work mainly focuses in spectrum selection process from the list of available licensed channels. It uses signal-to-noise ratio (SNR) as a parameter to generate the list of available licensed channels. Two algorithms have been proposed for selection of spectrum, namely random selection and game theory-based selection. In random selection, the channels are selected randomly and in case of game theory-based selection, a mixed strategy game is formulated. The payoff for the game is calculated as a function of channel holding time. The protocol is basically designed for environmental monitoring applications. Through the simulation results, author claims to have elongated the lifetime of the CRWSN, using both the algorithms. Also the game-based algorithm performs slightly better than random algorithm.

An auction-based GT approach has been used for heterogeneous WSN [19]. This method utilizes Stackelberg game technique to perform spectrum sharing. In this approach, the base station (BS) gathers the PU's spectrum details and auctions these frequency bands to the actor nodes which are then further disseminated among the contending nodes. The experimental results show that the bandwidth that has been purchased by actor nodes deployed in the heterogeneous network: (a) increases with the increase in the node distribution density and (b) is proportional to the distance between BS and the actor nodes. These results show that the reduction in interference enables the actor node to offload additional nodes for the licensed network. This happens at the cost of additional bandwidth consumed. The proposed scheme has been shown to outperform the available Stackelberg game-based approach for the static scheme, by achieving considerable increase the victim nodes' data rate. However, the scheme is not able to achieve a significant increase in the overall throughput, in comparison to that with the non-CR mechanism. The mechanism of spectrum sharing assumes uniform distribution of nodes in the underlying network.

A framework for jointly modeling spectrum sensing and sharing using cooperative game technique has been proposed in [20]. It uses Vickrey–Clarke–Groves (VCG) auction for allocating channel to SUs. There is a fusion center that manages the sensing and access policy for SU. The framework focuses on fairness, stability, and cooperation among the SU but energy consumption is not considered while designing the game.

An opportunistic channel assignment technique for cognitive nodes (CN) using GT approach has been proposed in [21]. A repeated game model called win-shift lose randomize (WSLR) strategy has been proposed. This approach maximizes the total network payoff and also ensures fairness to the cognitive nodes. The PU statistics is assumed to be known to all CN and work under limited memory and imperfect observations.

A work on reducing energy consumption in CRWSN using GT is given in [22]. It leverages on the capability of changing transmission parameters based on the sensed environment. The game is modeled as non-zero sum game. All SUs and PUs are the players as it focuses on improving energy consumption. The payoff function is derived as a function of energy, consumed with a change in communication channel. The simulation results show that using GT approach, the energy consumption is reduced by 30%.

The research work in CRWSN involving GT approach is still in a budding phase. Most of the work found in literature focuses on spectrum sharing using GT techniques in CRN.

3 Proposed Framework

CRWSNs are still in infancy stage and lot of research works are being carried out in this area. In this paper, we propose a framework to show how GT can be used with CRWSN to reduce energy consumption and to improve reliability in data delivery. The proposed framework is shown in Fig. 3.

The first step is spectrum sensing, where all the unused licensed channels are sensed and based on some parameters like SNR, a list of available licensed channels is generated. Once the list of available channels is known, a game is designed to allocate the channels to the contending SUs. The channel allocation game is defined as:

Game $G = \{I, S, u\}$ where

- Player set I: Players are all the nodes in CRWSN. Generally, the SUs are the players.
  $I = \{1, 2, \ldots, N\}$ where N-no. of nodes.
- Action space S: contains all the actions of players

$$S = \{s_1, s_2, \cdots, s_I\}$$
$$s_i\text{—action of node } i$$
$$s_{-i}\text{—action of other nodes except } i$$

- Utility Function u: for every node i, the utility function is denoted by $u_i(s)$.

The action or strategy of the SU can be framed depending on any one or more of the following objectives:

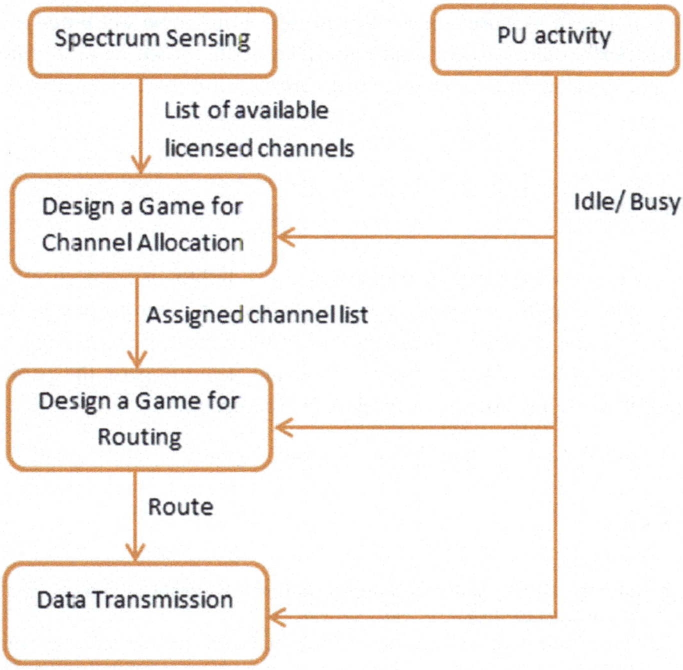

Fig. 3 Proposed framework

- To change the channel or not to change,
- Requiring some sensing order pattern,
- To keep the interference level below some threshold.

Similarly, the payoff or utility can be formulated as a function of any one or more of the following:

- Energy expenses associated with changing or not changing the channel,
- Probability of PU being present in the channel,
- Interference level,
- Residual energy.

Here, channel allocation as well as routing protocols are game theory based.

4 Implementation Plan

The proposed framework can be implemented after design and performing rigorous test of the channel allocation and routing protocols. In order to test the performance of the channel allocation and the routing protocol, a test plan has been prepared. Using the developed mathematical framework and set of mathematical expressions, the proposed channel allocation and routing protocols are to be simulated using MATLAB. Various performance evaluation parameters considered as mentioned below: energy efficiency, throughput, spectrum utilization, reliability, and network lifetime.

5 Conclusion

In this paper, a game theory-based framework for reliable and energy-efficient data delivery in cognitive radio wireless sensor network has been proposed. In the proposed framework, the channel allocation and routing protocols are suggested to be based on the principles of game theory. Detailed designing of the protocols and implementation work have been undertaken by the group.

References

1. Paul K, Avoidance I (2002) Spectrum policy task force. Federal Commun Comm Wash DC, Rep ET Docket 40(4):147–158
2. Mitola J (2000) Cognitive radio—an integrated agent architecture for software defined radio
3. Haykin S (2005) Cognitive radio: brain-empowered wireless communications. IEEE J Sel Areas Commun 23(2):201–220
4. Akan OB, Karli OB, Ergul O (2009) Cognitive radio sensor networks. IEEE Netw 23(4):34–40
5. Zahmati, AS et al (2009) Cognitive wireless sensor networks: emerging topics and recent challenges. In: 2009 IEEE Toronto international conference science and technology for humanity (TIC-STH). IEEE
6. Vijay G, Bdira EBA, Ibnkahla M (2011) Cognition in wireless sensor networks: a perspective. IEEE Sens J 11(3):582–592
7. Singhal D, Barjatiya S, Ramamurthy G (2011) A novel network architecture for cognitive wireless sensor network. In: 2011 international conference on signal processing, communication, computing and networking technologies. IEEE
8. Joshi G, Nam S, Kim S (2013) Cognitive radio wireless sensor networks: applications, challenges and research trends. Sensors 13(9):11196–11228
9. Morgenstern O, Von Neumann J (1953). Theory of games and economic behavior. Princeton University Press, Princeton
10. Rai R, Rai P (2019) Survey on energy-efficient routing protocols in wireless sensor networks using game theory. In: Advances in communication, cloud, and big data. Springer, Singapore, pp 1–9
11. Shi H-Y et al (2012) Game theory for wireless sensor networks: a survey. Sensors 12(7):9055–9097
12. AlSkaif T, Zapata MG, Bellalta B (2015) Game theory for energy efficiency in wireless sensor networks: latest trends. J Netw Comput Appl 54:33–61
13. Wang B, Wu Y, Liu KJR (2010) Game theory for cognitive radio networks: an overview. Comput Netw 54(14):2537–2561
14. Peizhe L et al A game-theoretic and energy-efficient algorithm in an improved software-defined wireless sensor network. IEEE Access 5:13430–13445
15. Hao X-C et al (2015) Energy efficient based channel assignment game algorithm for wireless sensor network. Wirel Pers Commun 85(4):2749–2771
16. Bacci G et al Game theory for networks: a tutorial on game-theoretic tools for emerging signal processing applications. IEEE Sig Process Mag 33(1):94–119

17. Belghiti ID et al (2016) Coalitional game-based behavior analysis for spectrum access in cognitive radios. Wirel Commun Mobile Comput 16(14):1910–1921
18. Salim S, Moh S (2016) An energy-efficient game-theory-based spectrum decision scheme for cognitive radio sensor networks. Sensors 16(7):1009
19. Sun S et al (2016) A Stackelberg game spectrum sharing scheme in cognitive radio-based heterogeneous wireless sensor networks. Sig Process 126:18–26
20. Rajasekharan J, Koivunen V (2015) Cooperative game-theoretic approach to spectrum sharing in cognitive radios. Sig Process 106:15–29
21. Khan Z et al (2016) Opportunistic channel selection by cognitive wireless nodes under imperfect observations and limited memory: a repeated game model. IEEE Trans Mobile Comput 15(1):173–187
22. Romero E et al (2014) A game theory based strategy for reducing energy consumption in cognitive WSN. Int J Distrib Sens Netw 10(1):965495

Routing Techniques in Internet of Things: A Review

Amlan Jyoti Dey and Hiren Kumar Deva Sarma

Abstract In recent time, Internet of Things (IoT) is one of the most popular subjects in various fields of development be it in academia or in industry. Machine learning can help in deriving knowledge from within the data available from millions of interconnected machines or devices. IoT networks are self-organized and decentralized, and such networks experience dynamic changes in their topology and node position. Hence, routing plays a very important role when data packets are transmitted between objects or things. Successful delivery of the data packets is to be ensured. Several nodes in IoT environment may undergo constant movements which result into breaking of interconnections between the devices. The change of topology that too frequently, is unavoidable in IoT networks. Routing of the data packets with efficiency and reliability is a great challenge that needs urgent attention. There has been effort in devising machine learning-based routing techniques for IoT environment. In this article, a survey on different routing techniques developed for IoT environment has been presented. Future research directions in the field of IoT have been outlined.

Keywords Internet of Things · Machine learning · Routing protocol · Topology

1 Introduction

The usage of Internet is spreading day by day to every spheres of human life starting from Internet surfing, electronic mails, playing games, research, electronic news, shopping, etc., to the most talked about concept inn recent time, Internet of Things (IoT). With the maturity of Internet, this new concept has emerged which has been known popularly as IoT (Internet of Things). So far, the available literature is concerned [1–3], the term "Internet of Things" has been proposed by Mr. Kavin Ashton.

A. J. Dey
NIIT Technologies Ltd., SEZ Greater Noida, Greater Noida 201308, India

H. K. D. Sarma (✉)
Sikkim Manipal Institute of Technology, Sikkim Manipal University, Majitar, Sikkim 737136, India
e-mail: hirenkdsarma@gmail.com

H. K. D. Sarma et al. (eds.), *Trends in Communication, Cloud, and Big Data*, Lecture Notes in Networks and Systems 99, https://doi.org/10.1007/978-981-15-1624-5_5

In a nutshell, IoT is the concept of connecting any device to the Internet. IoT also enables connecting any device to other connected devices. Thus, IoT leads to a huge network of connected things and people. Such a network collect and share data and also enables users to use these data for other purposes, as desired by the user. Therefore, huge data is available in IoT environment.

Devices and objects those are connected to Internet of Things networks have sensors built-in it, and the IoT platform integrates data across different devices. Data analytics techniques are applied to such data to discover the necessary information out of it. Finally, such information is shared with the applications which are designed to solve some specific problems or to address some particular needs.

Main objectives of Internet of Things ecosystem are summarized as mentioned below.

(i) IoT is to develop highly interconnected engineering system. Here, different devices equipped with necessary sensors are the main users.
(ii) IoT-based systems are expected to be intelligent enough to work for the sophistication of human lives.
(iii) IoT-based systems are also expected to improve the interrelationship between the humans and the living environment of human beings.

As per Forbes [4], the global market of Internet of Things-based systems will grow from $157 billion to $457 billion during 2016 to 2020. Industries from various sectors like manufacturing, logistics, and transportation are going to be the major players for the investments in this IoT market.

The following block diagram in Fig. 1 represents interlinks among various entities constituting IoT system.

The following are some of the real-life applications/deployment scenarios of IoT.

(a) Smart home technology

Home automation essentially offers homeowners all types of security and comfort along with convenience. Moreover, due to home automation, various devices or

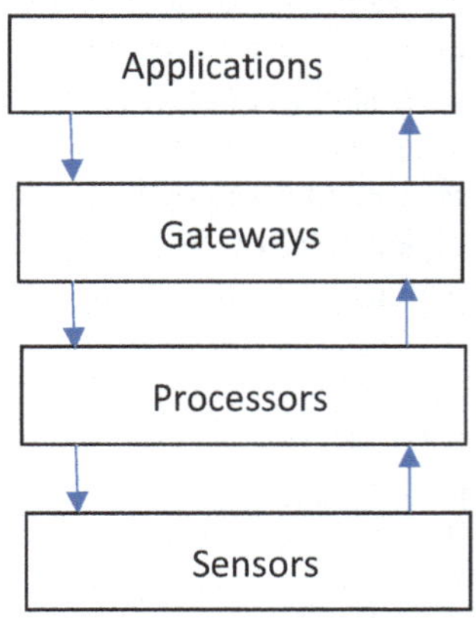

Fig. 1 Block diagram representing IoT system

systems consuming energy in a house shall efficiently utilize energy, due to proper control of these smart devices. Many organizations are working on this to build systems to make this dream a reality. As per [5], the funding in the line of home automation has already crossed $3 billion.

(b) Smart car

Another popular application of IoT is smart car [6]. Smart cars are expected to be driverless and such cars are equipped enough with technologies to drive itself to a defined destination. Smart cars are equipped with different devices and technologies such as sensors, cloud architecture, and Internet. The smart cars need to use huge volume of data related to traffic, pedestrians, various road conditions such speed breakers, potholes, corners, and sharp turns. Integrated processors process these data in real time. Various controllers or controlling units of the car use these outcomes of the processors and take different decisions for driving accordingly. The principles of subjects like artificial intelligence and machine learning shall have great role in making the dream of smart car a reality. Companies like Tesla, Volvo, BMW, Uber and Google are developing technologies for smart car.

(c) Amazon Go

Amazon has introduced the concept of IoT in retail stores where there will not be any cashier or cash counter [7]. Using sensors, online wallets and the customer's bank account, computers and machines will enable shopping experience for customers.

(d) Smart City

Cities are growing very rapidly and this growth is becoming a major challenge for the people who live in the cities in terms of healthy living and safety environment. According to Gartner [8], by 2050, it is expected that approximately 70% of the world's population shall be migrated towards the cities and live in cities. To accommodate this new demand on cities, IoT-based smart city concept shall play a very important role. While working for smart city concept, Nokia has launched "IoT for Smart Cities". It may be possible to deliver and manage various smart city services by such a framework. This framework is modular and scalable however integrated [9, 10].

As it is already mentioned that IoT is a networked environment, it is straightforward that various network protocols are to be designed and implemented in order to make IoT a reality. Similarly, network layer protocols shall play a very important role in routing of data packets meeting the design objectives. There have been different routing protocols already developed for IoT environment. In this article, various routing protocols applicable on IoT environment that have been designed using machine learning concepts, are discussed. Rests of the article are organized as mentioned below. Section 2 highlights importance of routing in Internet of Things, along with applicability of machine learning concepts in routing. Section 3 outlines various challenges faced in designing routing protocols for IoT, followed by Sect. 4, in which a survey of various machine learning-based routing protocols has been presented. Section 5 highlights various open issues related to routing in IoT. The article is concluded in Sect. 6.

2 Importance of Routing in IoT

Routing plays a vital role in IoT-networked environment. Routing protocols specify how the devices in the network communicate with each other in the network. Control information also known as control packets is circulated to select the best routes between any two nodes among multiple routes. Routing protocols determine the route for data packets to traverse in order to get delivered at the destination node (also known as sink) originating at a source node. Based on the algorithms used in routing, it decides the best path between the source and the destination node. Different researchers have suggested different algorithms and protocols in order to increase the network lifetime. Thus, energy efficiency is an objective sometimes in routing.

Machine learning concepts can also be implemented in IoT especially to improve the efficiency of routing. Machine learning (ML) is another important component of artificial intelligence (AI). ML enables devices to learn automatically and improve its performance from experience that too without being programmed explicitly.

Recently, machine learning techniques have been found being used extensively for varieties of computing tasks. These techniques for classification, regression, and density estimation are applied in different application areas such as bioinformatics, medical image processing, speech processing, network security, computer vision, cyber security and cyber forensic, and computer networks. The concepts behind the algorithms and techniques are derived from various fields that include mathematics, statistics, bio-science, and computing science.

3 Challenges in Routing in IoT

Internet of Things is a typical environment in which routing is a challenging task for various factors. Some of those factors are enlisted below.

Devices: Devices in the IoT environment may be of similar nature or of heterogeneous nature. Thus, heterogeneity present in the devices leads to a complex situation when interoperability is an issue.

Manufactures: The manufacturers of the devices used in IoT ecosystem may be the same one or different which may create differences, and it leads to a complex situation, while data transfer among the devices is a concern.

Network: The source node wherefrom that data gets generated and the destination node for which the data are destined may exist on the same or different networks. Thus, transferring data as well as control packets across the networks is again a challenge.

Connectivity: The connectivity between any two devices in the IoT network may or may not be stable. This depends on the network and the environment. Efficient transfer of data in the presence of faults in the networks is a challenge.

Resources: Insufficient resources namely available bandwidth, processing and storage capability of the devices, and battery power of the devices make routing a challenging task.

Cooperation: Cooperation among the devices in data relaying is desired, however, to achieve this has been a challenging task. Devices may be non-cooperative mainly due to the constraints regarding availability of resources. Smooth operation considering such facts is a challenge.

Communication process: Communication process, i.e., single-hop or multi-hop, in routing is again a challenge considering dynamic IoT environment.

Network topology: Network topology is frequently changing and this is due to mobile devices and other resource constraints like battery power/energy in the devices, pose challenges in routing in IoT environment.

Communication range: Variety of communication ranges are available among different devices which are manufactured by different vendors. This fact leads to a complex situation while considering connectivity between the devices.

Harsh environmental conditions: Heavy rain, storm, high temperature, etc., are some harsh environmental conditions which may lead to the malfunctioning of the IoT devices. This may lead eventually towards network failure.

D2D addressing mechanisms: Unavailability of a universally acceptable and unique addressing mechanism for the IoT devices is a challenge. If this gap is bridged that Device to Device (D2D) communication shall become easier.

4 Survey on Some Existing Routing Protocols of IoT

In this section, various routing protocols developed for Internet of Things (IoT) have been discussed. These are as mentioned below.

A. Naive Routing

The idea deployed in naive routing [11] is widely accepted. Each node can overhear its neighbours within its range. The source node sends through the network with route request packets called as beacons. Accordingly, destination nodes respond with a route reply message to the beacon and communication link is established between these nodes. Beaconing is typically utilized for various purposes such as location tracking, discovering routes to destinations and tracking neighbours through keep-alive requests. Beacon interval is an important factor that can affect the performance of the route discovery process. A too small beacon interval may generate a huge number of beacons. On the other hand, a larger beacon interval may incur a lesser number of generated beacons. The routing protocols such as DSR [12], DSDV [13] and AODV [14] fall under this category. However, this flooding causes overhead in the network. By using steady-state transition probability, we can derive the power consumption for each node. Such a probability may give a clear picture about the individual node behaviour and the overall network behaviour.

B. Hierarchical Routing

Clusters of nodes are used in hierarchical routing. Different nodes form clusters as per different clustering algorithms. Polling may be used for clustering purpose. The cluster head administers the communications on behalf of the members of the cluster. Cluster head helps in achieving Group mobility by following some metric to devise the mobility pattern of the nodes in the cluster [15–18].

C. Query-based Routing

The principle of query-based routing is data dissemination within the network. A querying node can fetch data from any node in the network. Common examples are SPIN [16–19] and Directed Diffusion [20–23].

D. Multipath routing

The multi-path routing protocols intends to use alternate paths towards every destination. This results in distribution of the cost of forwarding packets among more nodes. Such an approach save energy consumption of individual nodes [16–19].

E. Probabilistic routing

Decision regarding routing is based on a probability value. Earlier this value was computed by a technique called gossiping. Data packets are flooded into the network like a rumour with a probability p. Unlike other flooding mechanisms, these packets are forwarded only once thus reducing traffic overhead. A highly structured approach is to refer the prior history of packet delivery and mobility pattern, based on this we can decide which nodes can form a route to the destination [17, 18].

F. On-Demand Distance Vector (AODV)

AODV computes path on demand. It is a reactive protocol. The single path discovered is loop-free. The source may discover multiple paths but finally, keeps the best path by discarding remaining paths. Any interruption in the route leads to packet drop as there is no alternate paths available. Overall packet delivery ratio is less due to such packet drops. If a route fails, a new path needs to be discovered beginning the entire process from and therefore, AODV consumes more network resources and gives rise to extra overhead [14].

G. On-Demand Multipath Distance Vector (AOMDV)

On-Demand Multipath Distance Vector (AOMDV) discovers multiple paths between a pair of source and destination. It reduces network overhead under high network dynamics. Discovered multiple paths are loop free and also disjoint. Efficient delivery of data and fault tolerance are the benefits of this protocol. In the event of failure during routing, alternate path may be chosen from within the pool of discovered paths. If no alternate path is available then a route error message is produced. When a paths is chosen with a large number of intermediate hops, then probability for the occurrence of fading is high [24].

H. RPL

RPL [25] is based on IPv6 technology. It is a link independent routing protocol. RPL builds a robust topology over loose links. This routing protocol achieves multipoint-to-point, point-to-multipoint and point-to-point communication for different types of complex traffic models. RPL core is represented by a Destination Oriented Directed Acyclic Graph (DODAG), which is a directed acyclic graph with a single root. Each node knows its parent but has no knowledge about its related children. RPL maintains at least one path for each node to the root node. This path is generally through a preferred parent. RPL strives for faster path establishment and overall improvement in network performance.

5 Open Issues Related to Routing in IoT

Communication technology has achieved tremendous advancements. Social networks are latest additions to Internet like platforms. Considering these facts it is expected that a new generation of Internet (also called as future Internet) may appear in the years to come. There are several issues to be resolved in order to have IoT systems in place as expected. These challenges and open issues are to be addressed by the research community. It is also expected that existing architecture of Internet may require some reforms.

Some open issues in IoT are mentioned below.

Security Security provisioning in IoT is going to be difficult along with the increased level of automation of the devices. And such types of automations are creating new security issues. Cyber security has a great challenge today and having a defence mechanism against this is still an open issue.

Data management As the communications between the devices are being carried out, huge amount of data are getting generated every day between the devices. A large amount of data are to be transferred from one node to another. We need to ensure whether the intended data are being transported or not. Data management techniques are going to play a very important role in IoT.

Storage management As there are generations of a huge amount of data by the devices, subsequently data storage is a challenge. When multimedia data are considered the storage requirements are even more. Storing such data in the system and then accessing those efficiently are two big issues.

Server technologies Along with the growth in the number of devices in the network, overall number of requests and responses of the devices also increase. Managing such huge number of messages obviously depends also on the server technology, where the interfaces for communication are being run. It is expected that response of the server shall be quick against the request of any device in the network. No delay for presenting the response to the client is expected. So, sever technologies should be fully upgraded to the environment. And this is an ever-changing requirement.

Insecure authentication/authorization Authentication essentially means issuing right or permission for a user to access the information. Similarly, authorization means issuing right or permission to the user to edit or change the data. Ensuring security in these aspects, e.g., authentication and authorization, is an open issue since the IoT is again an open environment indicating that communication happens over insecure channels.

Interoperability As there are large numbers of heterogeneous devices operating in different platforms or technologies in the entire IoT environment, the devices should be effectively handled. So, there is a need for having a universally acceptable mechanism for interoperation of the underlying technologies. End-to-end interoperability is an open issue in IoT.

Mobility Mobility of IoT devices is a challenge for implementing IoT systems. In IoT-based systems, communication needs to be established with mobile users and services should be continuous without any interruption even when the devices are in move. Mobility of the devices may lead towards intermittent communication links. There may be interruption when information transfers from one gateway to another. Successful communication can be achieved by caching and proper channelization of the services, which allow applications to access data even when resources are temporarily unavailable. Ensuring successful communication in a mobile and so dynamic environment is an open issue.

Energy Energy is a limited resource for the devices in the IoT environment. Thus, ensuring successful communication between the devices in IoT environment in an energy-efficient manner is an open issue.

6 Conclusion

Internet of Things (IoT) has emerged as one of the most popular technologies in recent time. There are several aspects in which researchers need to focus in order to have correct implementation of IoT technology. These aspects cover a wide range starting from the development of efficient algorithms for different layers of the protocol stack to ensuring security to the users. In this article, importance of routing in IoT environment has been discussed. Several routing approaches have been outlined. Energy efficiency is one of the key concerns in IoT ecosystem. Therefore, energy-efficient routing in IoT is an important issue. Different open issues in IoT environment have been enlisted which also provides future directions for research in this field.

References

1. Gubbi J, Buyya R, Marusic S, Palaniswami M (2013) Internet of Things (IoT): a vision, architectural elements, and future directions. Futur Gener Comput Syst 29(7):1645–1660
2. Montenegro C, Rodríguez J (2015) Redes de Ingeniería Redes de Ingeniería Redes de Ingeniería, pp 37–44
3. Weber RH (2009) Internet of things—need for a new legal environment? Comput Law Secur Rev 25(6):522–527
4. Internet of Things by the number: market estimated and forecast. Forbes (2015). http://www.forbes.com/sites/gilpress/2014/08/22/internet-of-things-by-the-numbers-marketestimatesand-forecasts/
5. Malche T, Maheshwary P (2017) Internet of Things (IoT) for building smart home system. In: International conference on I-SMAC (IoT in social, mobile, analytics and cloud) (I-SMAC), at Palladam, India
6. Bajaj RK, Rao M, Agrawal H (2018) Internet of things (IoT) in the smart automotive sector: a review. In: IOSR Journal of Computer Engineering (IOSR-JCE), conference on recent trends in computer engineering (CRTCE), pp 36–44
7. https://internetofbusiness.com/amazon-go-eliminate-queues/
8. Gartner: Gartner says 4.9 billion connected 'Things' Will Be in Use in 2015 (2014)
9. Hofer S (2014) Smart cities and the internet of things, pp 485–494
10. Barriga KD, Romero CDG, Molano JIR (2016) Proposal of a standard architecture of IoT for smart cities. In: Learning technology for education in cloud the changing face of education, pp 77–89
11. Jeba N, Kamala V (2016) A survey on routing protocols for internet of things. Int J Adv Res Sci Eng Technol 3(5):1993–1996
12. Johnson D, Maltz D, Jetcheva J (2002) The dynamic source routing protocol for mobile ad hoc networks. Internet Draft, draft-ietf-manet-dsr-07.txt
13. Perkins CE, Watson TJ (1994) Highly dynamic destination sequenced distance vector routing (DSDV) for mobile computers. In: ACM SIGCOMM'94 conference on communications architectures, London, UK
14. Das S, Perkins C, Royer E (2002) Ad hoc on demand distance vector (AODV) routing. Internet Draft, draft-ietf-manet-aodv 11.txt
15. Handique M, Rai P, Biradar SR, Sarma HKD (2006) Energy efficient hierarchical cluster based communication protocol for wireless sensor networks with base station mobility. In: Proceedings of IEEE CODEC, Kolkata, India, pp 214–221
16. Singh SP, Sharma SC (2015) A survey on cluster based routing protocols in wireless sensor networks. Procedia Comput Sci 45:687–695 (2015)
17. Al-Karaki JN, Kamal AE (2004) Routing techniques in wireless sensor networks: a survey. IEEE Wirel Commun 11(6):6–28
18. Manap Z, Ali BM, Ng CK, Noordin NK, Sali A (2013) A review on hierarchical routing protocols for wireless sensor networks. Wirel Pers Commun 72(2):1077–1104
19. Zin S, Anuar NB, Kiah LM, Pathan A-SK (2014) Routing protocol design for secure WSN: review and open research issues. J Netw Comput Appl 41:517-530
20. Patil M, Biradar RC (2012) A survey on routing protocols in wireless sensor networks. In: 18th IEEE international conference on networks (ICON), pp 86–91
21. Younis O, Fahmy S (2004) HEED: a hybrid, energy-efficient, distributed clustering approach for ad hoc sensor networks. IEEE Trans Mob Comput 3:366–379
22. Sudeep T, Kumar N (2015) Joel JPC Rodrigues: a systematic review on heterogeneous routing protocols for wireless sensor network. J Netw Comput Appl 53:39–56
23. Intanagonwiwat C, Govindan R, Estrin D, Heidemann J, Silva F (2003) Directed diffusion for wireless sensor networking. IEEE/ACM Trans Netw 11(1):2–16

24. Tian Y, Hou R (2010) An improved AOMDV routing protocol for internet of things. In: International conference on computational intelligence and software engineering (CiSE), pp 1–4
25. Winter T, Thubert P, Brandt A, Hui J, Kelsey R, Levis P, Pister K, Struik R, Vasseur J, Alexander R (2012) RPL: IPv6 routing protocol for low-power and Lossy networks, IETF RFC 6550. Available in http://www.ietf.org/rfc/rfc6550.txt

Cloud

Cloud Detection and Cloud Removal of Satellite Image—A Case Study

Sanju Das, Purnendu Das and Bishwa Ranjan Roy

Abstract The book contributes to the subject area of remote sensing and Geographic Information System. It is focused on the study and analysis of automated cloud detection and removal of satellite imagery using the selection of thresholds value for various spectral tests in the perspective of RSGIS (Ramya, KarthiPrem, Nithyasri in IJIACS 3(2), [1], Rafael, Richard in Digital image processing. Prentice Hall, [2]). A significant obstacle of extracting information using satellite imagery is the presence of clouds. Removing these portions of image and then filling in the missing data is an important image-editing task. Traditionally, the objective is to cut the cloudy portions out from the frame and fill in the gaps with clear patches from similar images taken at different time. Remote sensing is providing opportunities in various branches of environmental research. The fields of application for multi-spectral remote sensing instruments in earth observation are monitoring the forests, oceans or urban areas over agricultural applications to the extent of natural resources. A significant prerequisite for analysis of earth observation data is the information that is free from external influences and disturbances. One possible cause of data loss is cloud cover of satellite imagery. Cloud cover is recognized as a significant loss of data and information quality by many scientific studies. The existence of cloud cover is the loss of meaningful data and information because they are a considerable source of uncertainty with regard to the application of any algorithm aiming for the retrieval of land surface (Zakaria, Ibrahim, Suandi in A review: image compensation techniques. pp. 404–408, [3], Sengee, Sengee, Choi in IEEE Trans Consum Electron 56(4):2727–2734, [4], Hardin, Jensen, Long, Remund in Testing two cloud removal algorithms for SSM/I [5]).

Keywords Remote sensing · Cloud detection · Cloud removal · Satellite imagery · Thresholds · Image reconstruction · Reference image · Cloud free image · Information extraction · Colour conversion · Pixel-to-pixel correction

---

S. Das (✉) · P. Das · B. R. Roy
Department of Computer Science, Assam University, Silchar, India
e-mail: sanjudasp@gmail.com

H. K. D. Sarma et al. (eds.), *Trends in Communication, Cloud, and Big Data*, Lecture Notes in Networks and Systems 99, https://doi.org/10.1007/978-981-15-1624-5_6

1 Introduction

A significant obstacle is the clouds in the satellite imagery while extracting information by satellite to a remotely sensed object on the earth. Various approaches have already developed with different result for detecting and removing clouds contaminated region from satellite imagery [6, 7].

Satellite images are useful for identifying the changes occur in the earth observation such as new settlement, change of river direction, land use for crop production and natural calorimetric [2, 4]. The major problem with these images is that the region below clouds cover is not covered properly by the satellite sensor. The image distortion occurs due to cloud cover of remotely sensed objects.

In this chapter, an automated cloud detection and cloud removal (ACDCR) algorithm has been proposed. Detection of cloud in satellite imagery is done based on the pixel intensity (PI) value, and cloud removal is done by merging the spatial details of one subject image and two reference image having same location. This process is applied for whole subject image to obtain a cloud-free image. To identify exact location of cloud contaminated region, cloud detection based on threshold value approach is introduced.

2 Thresholds

Threshold is one of the methods used for image segmentation. It is useful in discriminating subject image from the reference image. By selecting an adequate threshold value T, the grey-level image can be converted to binary image. The binary image should contain all of the essential information about the position and shape of the objects of interest of the subject image. The advantage of obtaining a binary image is that it reduces the complexity of the data and simplifies the process of recognition and classification. The most common way to convert a grey-level image to a binary image is to select a single threshold value T for whole image and apply the conditions for all the grey-level values:

$\leq T$ will be classified as black (0), and
$> T$ will be white (1).

The crux of the segmentation problem becomes one of selecting the proper value for the threshold T. In this case, the value of T is selected between cloudy and cloud-free modes.

3 Model of the Proposed Method

The objective is to identify clouds in a subject image and replace them from the reference images that are acquired at the different period of time. The workflow of the proposed technique for cloud detection and cloud removal from satellite imagery consists of the following steps: cloud detection, cloud removal and image reconstruction. In the first step, an automated detection method is applied to detect the cloudy pixels from the subject image. In the next step, the detected clouds are corrected and removed using the reference image. In the last step of the proposed model, image reconstruction is performed to fill out the missing information in the subject image and to produce a cloud-free image (Fig. 1).

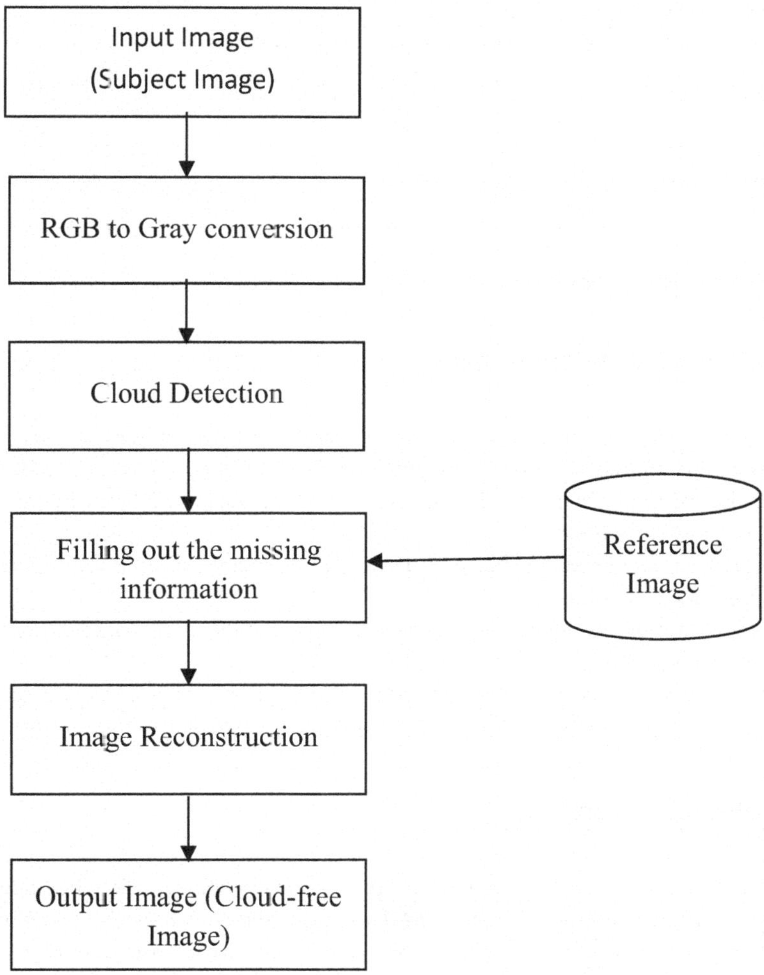

Fig. 1 Workflow of the proposed method

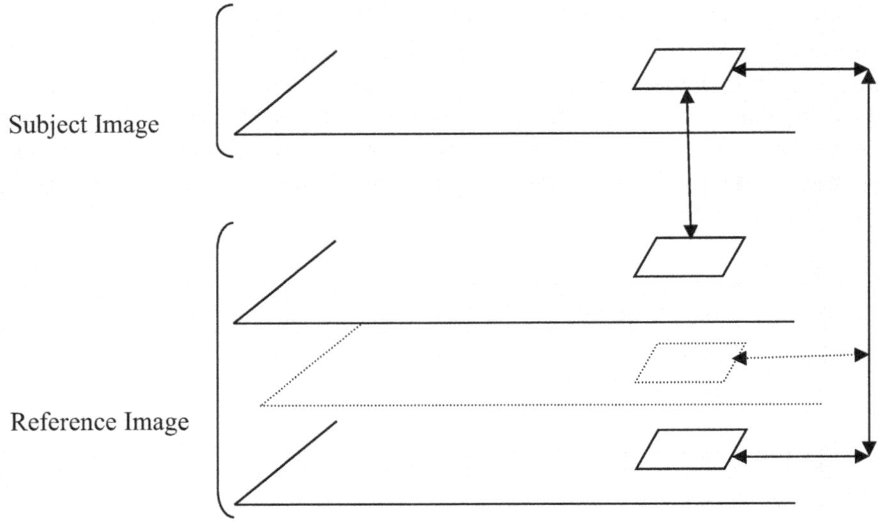

Fig. 2 Pixel-to-pixel comparison

4 Information Extractions

The aim is to extract correct information from earth by using satellite; various factors play an important role while extracting information by satellite. One of the factors is the presence of clouds in the satellite imagery. The following hypothesis is considered for the proposed method:

A significant increase of reflectance between two images of time series is related to the presence of clouds.

As the clouds are moving, clouds may not present at the same coordinates in two different ortho-photos.

Clouds are generally much brighter compared to other object in earth (Fig. 2).

5 Colour Conversion

To detect the clouds in the target image, the pixel intensity value needs to be calculated first. For the given red–green–blue (*RGB*) target image, it must be converted to any of the colour space for further processing. Here we choose the *YIQ* colour space conversion, where *Y* represents the luma information, *I* and *Q* represents the chrominance information. The *Y* channel alone is used to find the average intensity value. The *RGB* to *YIQ* colour conversion is based on the following formula:

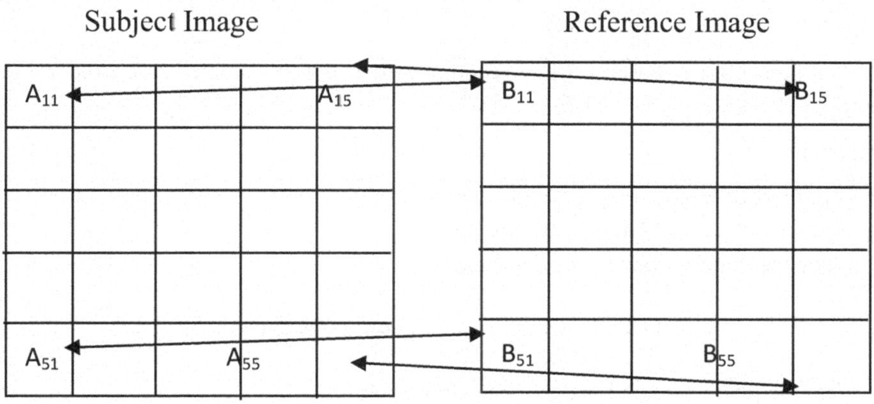

Fig. 3 Pixel-to-pixel image correction

$$= \begin{pmatrix} Y \\ I \\ Q \end{pmatrix} \begin{pmatrix} 0.229 & 0.587 & 0.114 \\ 0.595716 & -0.274453 & -0.321263 \\ 0.211456 & -0.522591 & 0.311135 \end{pmatrix} \begin{pmatrix} R \\ G \\ B \end{pmatrix}$$

The major advantage of *YIQ* is the greyscale information separated from colour data. So the same signal can be used for both colour and black-and-white images.

6 Pixel-to-Pixel Correction

Pixel-to-pixel mapping is the process of matching pixel of one image to another image so the same geographic area is positioned coincident with each other, respectively. This type of correction is required when any changes occurred or part of the original image gets affected due to the presence of some obstacles while satellite is capturing the image. A reference image needs to be used to get rid of the geometrically corrupted area. The following figure shows the two images of same location at different time series (Fig. 3).

7 Cloud Detection

The detection of clouds in satellite imagery in the proposed method is based on pixel intensity value. Initially, a threshold value is set which is known as pre-defined threshold value. If the pixel value in the target image is lower than the pre-defined threshold value, the pixel is marked as cloudy pixel. This process is repeated for each and every pixel in the target image.

Algorithm for Detecting Cloudy Pixel

The algorithm is based on the assumption that the pre-defined threshold value ranges between $T_i = 95$ and $T_i = 255$.

Algorithm:

if the DN of image > 95 & $\leq$ 255,
then the pixel is considered as cloudy pixel
place the cloudy pixel in the database.

$$\text{DB1} = \text{All the cloudy pixel in the image}$$

Else the pixel is considered as cloud-free pixel
place the cloud-free pixel in the database

$$\text{DB2} = \text{All the cloud -free pixel in the image.}$$

7.1 Procedure of Cloud Detection Algorithm

In the proposed work, initially one target image (T_i) and two reference images ($R1$, $R2$) of different time interval having same geometric coordinates have been considered. The target image is taken as input image, and the reference images ($R1, R2$) are referred from the database. The digital number (DN) values are compared between the target image $T\ (i, j)$ and the reference images $R1(i, j)$, $R2(i, j)$, respectively. If the DN of a particular pixel in (i, j) location of the target image is having lower than 95, it is considered as cloudy pixel. After finding the cloudy pixel, this particular pixel is replaced from the reference image having the minimum DN value. This process is repeated for all the pixels in the target image.

8 Cloud Removal

After detecting the cloudy pixel in the target image, it needs to be removed to get the information about the earth observation correctly. As the method is based on pixel-to-pixel comparison, the cloudy pixels are replaced and removed from the reference image by overwriting their spectral data from the database which was stored in the cloud detection process.

8.1 Algorithm for Cloud Removal

Step 1: Input the image as target image
Step 2: Extract spatial information from reference images ($R1$, $R2$)
Step 3: The cloudy pixel in the target image T (i, j) is compared with reference images $R1(i, j)$, $R2(i, j)$, respectively, to get the best DN value.
Step 4: The best DN value either from $R1(i, j)$ or $R2(i, j)$ will be replaced with T (i, j) using TF $= T\ (i, j) \parallel R1(i, j) \parallel R2(i, j)$
Step 5: Merge the spatial details of three images to produce a cloud-free image
Step 6: Repeat Steps 3–5 for all the pixels in the target image.

The only requirement of the algorithm is that images should be captured at different time series so that they have different cloud cover patterns at different geometric position.

The algorithm is quite simple to implement: first the images converted to greyscale images, and the intensity of pixel is calculated. Next, a threshold value $T_V = 95$ is chosen as pre-defined threshold value. Then, the threshold value is applied to the target image T_i to find the cloudy and cloud-free pixels. Finally, the detected cloudy pixel in the target image T_i is removed and replaced with the data from another images of same location. This process is repeated for all the pixels in the target image.

The proposed algorithm was tested on more than ten target images and their corresponding reference image. The data set of satellite image taken from san-jose-costa-rica-latin-america. The cloudy pixels are detected and removed correctly in the target image from the reference image using the proposed automated cloud detection and removal algorithm. Figure 4a shows the reference image, and Fig. 4b shows the cloudy subject image which is to be corrected with respect to the reference image (Figs. 5 and 6).

The detected cloud is removed and replaced with data from the image acquired at different time series for same location. To remove the cloudy pixel from the subject image, the new automated cloud detection and cloud removal (ACDCR) algorithm is applied. The final cloud-free image is shown in Fig. 7.

Greyscale image of (a) subject image, (b) reference image-1, and (c) reference image-2 is as follows: (Figs. 8 and 9).

9 Conclusion

The ACDCR algorithm described is an automatic method to detect cloudy pixel in the satellite subject image, and the detected cloudy region is removed and replaced by the reference image of same location. The missing information in the subject image is reconstructed from the reference image. As the approach is based on the pixel-to-pixel comparison, it produces a better image. This method was tested for various input image and their corresponding reference images.

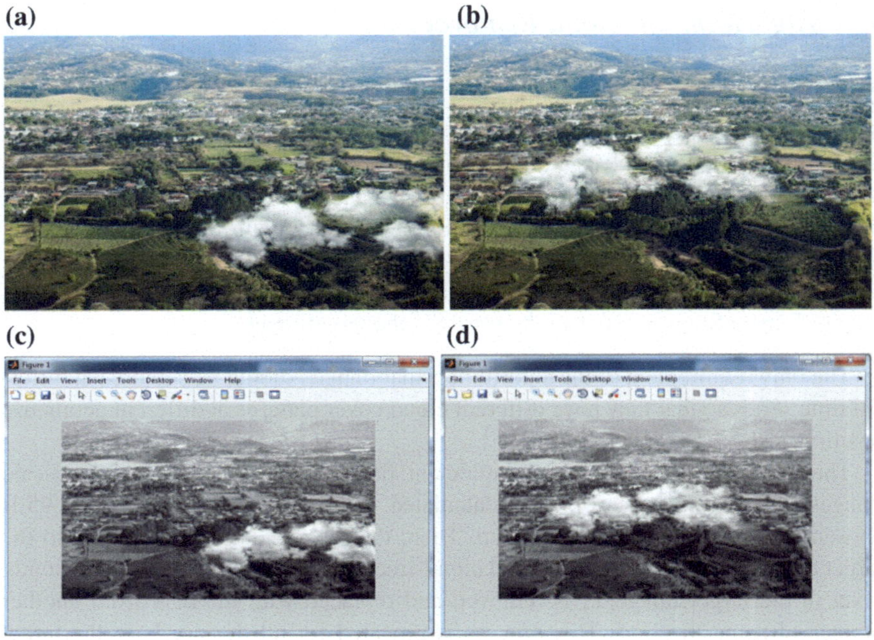

Fig. 4 **a** Subject image [8, 9], **b** reference image, **c** greyscale image of subject image and **d** greyscale image of reference image

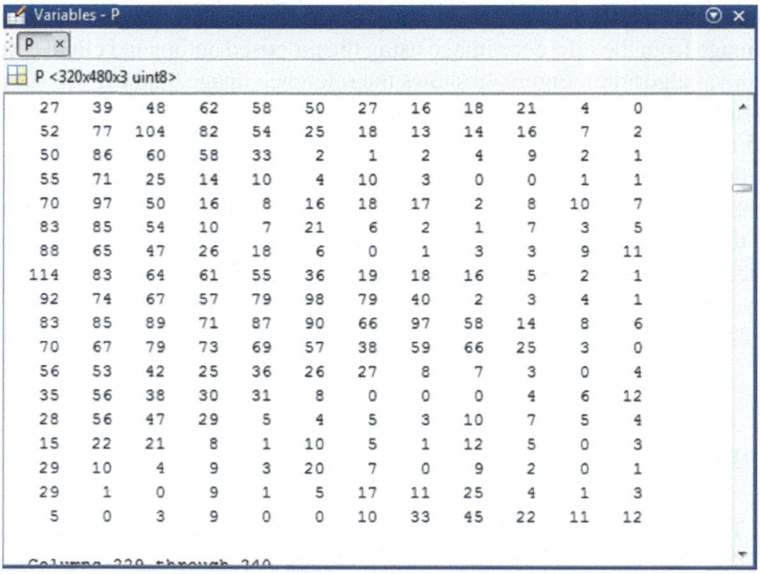

27	39	48	62	58	50	27	16	18	21	4	0
52	77	104	82	54	25	18	13	14	16	7	2
50	86	60	58	33	2	1	2	4	9	2	1
55	71	25	14	10	4	10	3	0	0	1	1
70	97	50	16	8	16	18	17	2	8	10	7
83	85	54	10	7	21	6	2	1	7	3	5
88	65	47	26	18	6	0	1	3	3	9	11
114	83	64	61	55	36	19	18	16	5	2	1
92	74	67	57	79	98	79	40	2	3	4	1
83	85	89	71	87	90	66	97	58	14	8	6
70	67	79	73	69	57	38	59	66	25	3	0
56	53	42	25	36	26	27	8	7	3	0	4
35	56	38	30	31	8	0	0	0	4	6	12
28	56	47	29	5	4	5	3	10	7	5	4
15	22	21	8	1	10	5	1	12	5	0	3
29	10	4	9	3	20	7	0	9	2	0	1
29	1	0	9	1	5	17	11	25	4	1	3
5	0	3	9	0	0	10	33	45	22	11	12

Fig. 5 Pixel intensity value of subject image

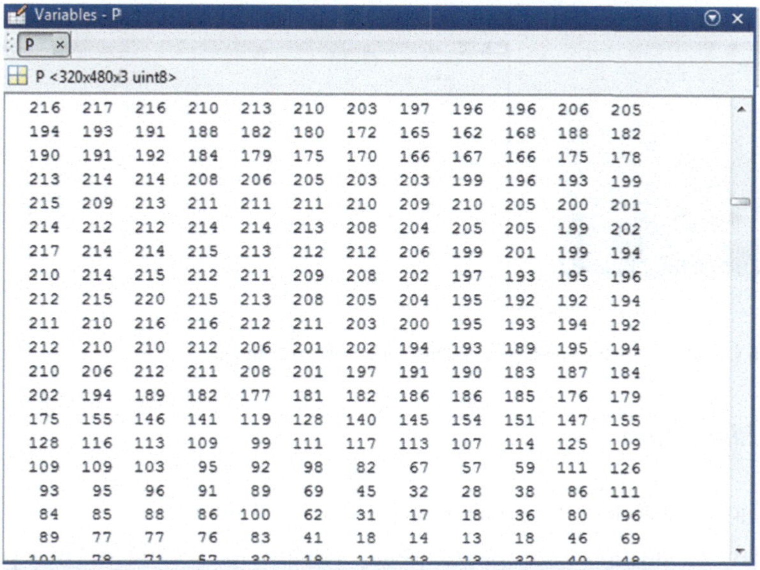

216	217	216	210	213	210	203	197	196	196	206	205
194	193	191	188	182	180	172	165	162	168	188	182
190	191	192	184	179	175	170	166	167	166	175	178
213	214	214	208	206	205	203	203	199	196	193	199
215	209	213	211	211	211	210	209	210	205	200	201
214	212	212	214	214	213	208	204	205	205	199	202
217	214	214	215	213	212	212	206	199	201	199	194
210	214	215	212	211	209	208	202	197	193	195	196
212	215	220	215	213	208	205	204	195	192	192	194
211	210	216	216	212	211	203	200	195	193	194	192
212	210	210	212	206	201	202	194	193	189	195	194
210	206	212	211	208	201	197	191	190	183	187	184
202	194	189	182	177	181	182	186	186	185	176	179
175	155	146	141	119	128	140	145	154	151	147	155
128	116	113	109	99	111	117	113	107	114	125	109
109	109	103	95	92	98	82	67	57	59	111	126
93	95	96	91	89	69	45	32	28	38	86	111
84	85	88	86	100	62	31	17	18	36	80	96
89	77	77	76	83	41	18	14	13	18	46	69

Fig. 6 Pixel intensity of reference image

(a) (b) (c)

Fig. 7 **a** Subject image, **b** reference image-1 and **c** reference image-2

(a) (b) (c)

Fig. 8 Greyscale image of **a** subject image, **b** reference image-1 and **c** reference image-2

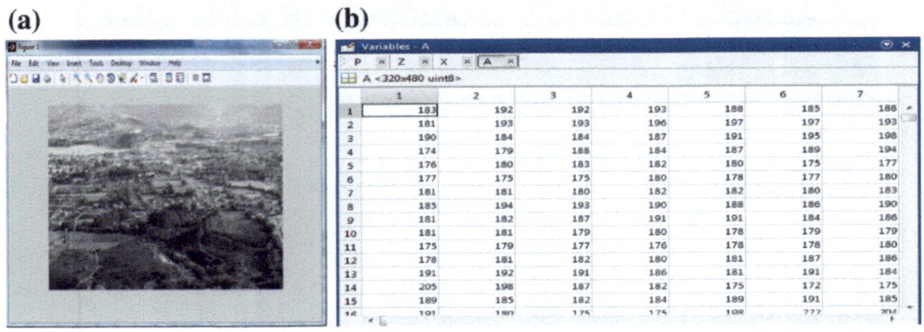

Fig. 9 **a** Cloud-free image and **b** pixel intensity of cloud-free image

9.1 Limitation

The major limitation of this research proposal is listed below:

- The live satellite images are not frequently available for a continuous time period.
- To work with large data set, it is required for a high-end machine.
- The varity of land, cloud is observed as the change in the general pattern. So it is difficult to set the threshold value for all the potential cases.

9.2 Future Work

- The proposed research is based on pixel-to-pixel evaluation on satellite imagery for the purpose of detection and correction of cloudy region in the images. The proposed approach works properly for large and thick cloud coverage area.
- The research may further be extended for identifying and removing thin and small cloud coverage area.
- The research may further be extended for identifying cloud shadow coverage area.
- The research may also be extended for classifying between clouds, and the object on the earth which appearance is same as cloud.

References

1. Ramya P, KarthiPrem S, Nithyasri A (2014) Cloud removal in high resolution satellite images using discrete wavelet transform. IJIACS 3(2). ISSN 2347–8616
2. Rafael CG, Richard EW (2002) Digital image processing, 2nd edn. Prentice Hall (2002)
3. Zakaria MF, Ibrahim H, Suandi SA (2010) A review: image compensation techniques. In: 2nd international conference on computer engineering and technology, vol 7, pp 404–408

4. Sengee N, Sengee A, Choi H-K (2010) Image contrast enhancement using bi-histogram equalization with neighborhood metrics. IEEE Trans Consum Electron 56(4):2727–2734 (2010)
5. Hardin PJ, Jensen RR, Long DG, Remund QP (1999) Testing two cloud removal algorithms for SSM/I
6. Teillet PM, Fedosejevs G (1995) On the dark target approach to atmospheric correction of remotely sensed data. Can J Remote Sens 21(4):374–387
7. Biday S, Bhosle U (2009) Relative radiometric correction of cloudy multitemporal satellite imagery. Int J Electr Comput Energ Electron Commun Eng 3(3):472–746
8. http://earthobservatory.nasa.gov/
9. http://news.satimagingcorp.com/

Big Data

Security Issues in Big Data

Hiren Kumar Deva Sarma

Abstract Big data analytics is one of the very popular technologies at the present moment. Ensuring security in the big data environment is a challenging task. Unique characteristics of big data environment make it very tricky to ensure security in it. There are unique challenges and requirements. Internet of Things and social networks are contributing in generating a huge amount of data and at the same time making the security scenario complex. In this article, various security threats, security challenges and security requirements in big data environment are reviewed.

Keywords Big data · Data analytics · Security · Threats

1 Introduction

In recent times, data have become a highly important asset not only for business organizations, but also for society as a whole. Starting from various types of industries to the government of every nation, they heavily rely on data for setting up of goals, making policies and taking decisions. In modern days, the entire society is driven by data. The amount of data generated every day in digital format, at present, is really huge. This is due to the proliferation of social networks, Internet of Things (IoT), mobile phones and availability of Internet even in remote corners of the world. Not only text but also multimedia data are generated at a high pace. This trend of generating huge volume of data every day is upward and ever increasing, and it does not appear that this trend shall be downward in the near future. As per [1], the amount of data generated in two days by humans is in the tune of five exabyte.

Data get generated from various sources including human and devices. These data are of heterogeneous nature as well as unstructured. Storage and processing of such type of data need novel ways and techniques, respectively. Such huge volume of data has potential and is highly useful for business organizations as well as governments, in the sense that by appropriate processing and analysis of the data one can get

H. K. D. Sarma (✉)
Department of Information Technology, Sikkim Manipal Institute of Technology, Sikkim Manipal University, Majitar, Sikkim 737136, India
e-mail: hirenkdsarma@gmail.com

H. K. D. Sarma et al. (eds.), *Trends in Communication, Cloud, and Big Data*, Lecture Notes in Networks and Systems 99, https://doi.org/10.1007/978-981-15-1624-5_7

insights into certain situations. Hidden information can be obtained by appropriate analysis of the data. However, traditional methods are not applicable for processing of such type of data with unique characteristics. This gives rise to a new field in the domain of data, namely big data analytics. Big data are characterized by five Vs, e.g. volume, variety, veracity, value and velocity.

Although there are various tools and platforms available for big data analytics, and organizations are using those, ensuring security in big data environment is a challenge. Novel techniques and tools are necessary for addressing security issues in the context of big data. Traditional security management approaches are not directly applicable considering the unique characteristics of big data [2]. The security goals in big data environments may be the same as computer networks or systems security goals, such as ensuring availability, confidentiality, authorization and authentication. But achieving those goals may not be possible through traditional tools and techniques.

In this article, an effort has been made to understand various security issues or threats in big data environment. Effort has been made to answer "Why is it challenging to ensure security in big data environment?" and various approaches/countermeasures already made available by the researchers to address the security issues in big data are highlighted.

Rest of this article has been organized as mentioned below. Section 2 provides an overview of big data environment. Section 3 outlines security fundamentals followed by Sect. 4 in which security threats in big data are outlined. Technical requirements for ensuring security in big data environment are mentioned in Sect. 5, and the article is concluded in Sect. 6.

2 Big Data Environment

Big data essentially refer to large volume of data which has become a reality in today's environment. Big data analytics strives to discover hidden knowledge otherwise buried inside big data. Handling of big data is not possible with traditional approaches used for databases. This is so because of its underlying complexities and unique features. Initially, three Vs, namely volume, velocity and variety were used to uniquely characterize big data [3]. Volume indicates the amount of data, velocity indicates the rate at which data are being generated, and variety indicates the types of data, e.g. structured, semi-structured and unstructured. Big data naturally refer to a very large amount of data, which is generated at a high rate, and also it is a combination of structured, semi-structured and unstructured data. At a later stage, two more Vs, namely veracity and value are added to characterize big data [4]. Veracity indicates uncertainty and impreciseness in data, and value refers to the fact that data under consideration are of great value to organizations or business houses or to any interested individual. In order to handle such a data pool, one needs a different way for modelling the data (i.e. data model), for storage, for processing and also for transmitting for various purposes. Not only the large business houses but also even small business houses are benefitted by the results of big data processing. Big data analytics can add great values to organization's decision-making processes, and also towards making strategies for survival in the competitive market. Even governments of different nation are exploiting the power of big data analytics and making use of it in respective governances.

Technologies like Apache Hadoop [5] enable organizations or users, for processing of big data. Hadoop is a distributed processing framework which can process big data across a cluster of computers using its own file system known as Hadoop Distributed File System (HDFS). Hadoop is an open-source implementation of MapReduce [6, 7] programming model.

However, security is an important dimension in big data ecosystem which was overlooked initially. Now, it has been realized, and many researchers are putting effort to make big data environment secure from all possible directions.

3 Security Fundamentals

Big data security objectives are same as that of other types of data. Authentication, confidentiality, integrity and availability are the security objectives for any kind of data [8, 9].

Why is it challenging to ensure security in big data environment?

Mere applications of cryptographic techniques may not be sufficient for handling security issues in big data. This is mainly due to the unique characteristics of big data as already described in terms of five Vs. Big data environment involves cloud computing paradigms, cloud storages, communication systems, social networks and Internet as a whole. Thus, security challenges faced by all the above-mentioned systems are also present in big data. Moreover, the distributed processing model of big data invites complex security challenges. Considering all the above parameters, ensuring security in big data is a challenging job.

4 Security Threats in Big Data

Cloud Secure Alliance (CSA), a non-profit organization aims to promote the uses of best practices for achieving security assurance in cloud environment [10]. Big Data Working Group is a part of CSA that focuses on ensuring big data services in a secure manner. CSA has identified four different dimensions of security and privacy challenges in big data ecosystem. These four dimensions are **Infrastructure Security**, **Data Privacy**, **Data Management** and **Integrity and Reactive Security**. Various security challenges faced by each of these dimensions are as mentioned below. Challenges under infrastructure security are secure distributed processing of data and security best actions for non-relational databases. Security challenges faced by data privacy dimension are data analysis through privacy preserving data mining, cryptography solutions for data security and granular access control. Different security challenges faced by data management are secure data storage and secure transaction logs, granular audits and data provenance. Challenges faced by integrity and reactive security dimension are end-to-end filtering and validation, supervising the security level in real time.

These security and privacy challenges cover the entire big data life cycle. Various constituting units of big data life cycle are sources of data production (i.e. devices of data origin), the data itself, data processing, data storage, data transport and data usage in different devices [11].

Internet of Things (IoT) is one of the recent developments in the world of Information and Communication Technology. As per Gartner report, 26 billion of IoT devices shall be deployed by 2020 [11]. These IoT devices are also sources of big data generation. IoT offers tremendous benefits and opportunities for the users of it; however, these offerings come with big security challenges. Hewlett Packard (HP) conducted a survey on security issues present in IoT solutions which are available in the market and concluded that 70% of such systems suffer from security problems. HP started Open Web Application Security Project (OWASP) in 2014 with an objective to identify the top ten security problems that IoT devices are going to suffer from and to suggest some solutions to those problems. Security problems identified in this study are as mentioned below.

(i) **Insecure web interface**: Attacker can get unauthorized access to control the IoT devices. Different security attacks may be cross-site scripting, cross-site request forgery, SQL injection, etc.
(ii) **Insufficient authentication and authorization mechanism**: This may allow attacker to break the passwords, and attacker may get access which is illegitimate.
(iii) **Insecure network services**: Such weaknesses may lead an attacker to access weak network services and use those services as jumping point for attacking other IoT devices in the network.
(iv) **No transport encryption**: Attacker can eavesdrop easily data in transit between IoT devices and support systems.
(v) **Privacy issues**: Users of IoT devices or services or support systems need to supply personal data; however, such devices or services or support systems fail to protect those data.
(vi) **Insecure mobile interface**: Attacker can access data in IoT devices or in IoT support systems and even can control the devices if the mobile interfaces do not have proper security control.
(vii) **Insecure cloud interface**: In the absence of proper security control in cloud interface, attacker may be successful in accessing data and controlling the devices.
(viii) **Insecure software or firmware**: Unencrypted data and unauthenticated connections may allow attackers to perform malicious update which in turn may compromise IoT devices or network of IoT devices.

(ix) **Insufficient security configurability**: Poor configuration mechanisms help attackers in getting data access and control on the device.

(x) **Poor physical security**: If any IoT device is physically accessible to attacker, then attacker can play with its operating system by accessing open ports like USB, and subsequently may get data access, and achieve control over the device.

5 Technical Requirements in Order to Ensure Security in Big Data

Perhaps, it is not possible to have one single solution to address all security issues in big data ecosystem. Security and privacy challenges are there in every stage of the life cycle of big data starting from origin of big data to the usage of it [11]. Various constituting units of big data life cycle are origin of data (i.e. the sources from where data get generated), data itself, data storages, data processing frameworks in general and processing nodes in specific, data transport and finally, the devices at which processed results get used (i.e. devices at the end user level). Security needs to be ensured at each unit of big data life cycle. Various security points from technology point of view are enlisted below.

Technology for maintaining data privacy: Maintaining privacy of data is important and for this cryptographic technique with enhancements might be adopted.
Technology for access control: Access to the big data ecosystem needs to be restricted. Access mechanisms need to be developed considering the requirements of big data environment.
Technology for secure storage of data: Storage of data, itself is a vital issue in big data environment. Ensuring security to such storages is a concern and needs to be met.
Technology for ensuring data integrity: In big data environment, data come from various sources which are not possible to be seen by anyone in the system. Thus, the origin of data is not possible to be seen. However, it is to be ensured that data originated from the desired sources and not from some other sources. It is a kind of integrity requirement. Technology to ensure such integrity is desired.
Technology for attack detection and recovery in real time: Attack may happen in the big data ecosystem at any time. Big data ecosystem covers a wide range of data sources and also application points. Detecting attack in real time is important, and also techniques for recovering from such attack in real time are necessary. Controlling further damage by recovering from such attack in the big data ecosystem is an important dimension to be addressed.

6 Conclusion

Security in big data is an important issue. Efficient security solutions are required for ensuring security to the end-users in big data environment. The challenges in big data security are unique due to unique characteristics of big data. However, the security requirements are the same as those of other types of data. Internet of Things and social networks have given rise to many security challenges. Technologies are required to ensure security in big data storage, to have access control and also to maintain data privacy. Traditional cryptographic techniques may not suffice here. Similarly, technologies to verify the origin of data and also to detect attack and recover from it in real time are required.

References

1. Sagiroglu S, Sinanc D (2013) Big data: a review. In: Proceedings of the international conference on collaboration technologies and systems (CTS), San Diego, CA, USA, 20–24 May 2013, pp 42–47
2. Demchenko Y, Ngo C, de Laat C, Membrey P, Gordijenko D (2014) Big security for big data: addressing security challenges for the big data infrastructure. In: Jonker W, Petković M (eds) Secure data management. Springer International Publishing, pp 76–94
3. Chen M, Mao S, Liu Y (2014) Big data: a survey. Mob Netw Appl 19:171–209
4. Khan MA-U-D, Uddin MF, Gupta N (2014) Seven V's of big data understanding big data to extract value. In: Proceedings of the 2014 zone 1 conference of the American Society for Engineering (ASEE zone 1 2014), Bridgeport, CT, USA, 3–5 Apr 2014
5. Cackett D (2013) Information management and big data a reference architecture. Oracle, Redwood City, CA, USA
6. Jiang D, Ooi BC, Shi L, Wu S (2010) The performance of MapReduce: an in-depth study. Proc VLDB Endow 3:472–483
7. Dean J, Ghemawat S (2004) MapReduce: simplified data processing on large clusters. Commun ACM 51:107–113
8. Stallings W (2006) Cryptography and network security, principles and practice, 5th edn. Pearson Education, Inc., publishing as Prentice Hall. ISBN 13: 978-0-13-609704-4
9. Moreno J, Serrano MA, Fernández-Medina E (2016) Main issues in big data security. Future Internet 8(44):1–16
10. Big Data Working Group, Cloud Security Alliance (CSA) (2013) Expanded top ten big data security and privacy, Apr 2013. Available online: https://downloads.cloudsecurityalliance.org/initiatives/bdwg/Expanded_Top_Ten_Big_Data_Security_and_Privacy_Challenges.pdf
11. Moura J, Serrão C (2016) Security and privacy issues of big data. In: Handbook of research on trends and future directions in big data and web intelligence, pp 20–52

Computation

Earth Mover's Distance-Based Automated Geometric Visualization/Classification of Electrocardiogram Signals

Kumar Dron Shrivastav, Priyadarshini Arambam, Ankan Mukherjee Das, Shazina Saeed, Upendra Kaul, Priya Ranjan and Rajiv Janardhanan

Abstract The tremendous load on the rather archaic medical system in developing countries has necessitated the need to implement artificial intelligence-enabled automated systems to classify different kinds of electrocardiogram (ECG) traces. To this end, we are proposing a novel R-based open-source software with inherent capability to classify different kinds of automated geometric visualizations along with its categorization based upon similarity indices as measured by earth mover's distance (EMD). This innovative automated software needs verification and validation by clinical practitioners/cardiologists before being implemented to classify large ECG databases to enhance its machine learning capabilities. We anticipate that integration of this robust automated classifier with divergent platforms such as mobile health applications would enable the subjects/patients to continuously monitor the heart rate themselves.

Keywords Cardiovascular diseases (CVDs) · ECG · EMD · Geometric visualizations

1 Introduction

Electrocardiogram (ECG) is a scientific representation of the electrical activity of the heart muscles in analog form over a period of time. Rising incidence of cardiovascular diseases (CVDs) presents a major public health challenge not only in India but elsewhere in the world [1]. Electrocardiogram is generally printed on a graph paper for analysis and diagnosis of CVDs [2]. The ECG traces are an output of the electrical activity in heart, when amplified and recorded in the form of analog signals for a

---

K. D. Shrivastav · P. Arambam · A. M. Das · S. Saeed · R. Janardhanan (✉)
Amity Institute of Public Health, Amity University, Noida, Uttar Pradesh, India
e-mail: rjanardhanan@amity.edu

P. Arambam · U. Kaul
Batra Hospital & Medical Research Centre, New Delhi, India

P. Ranjan
Amity School of Engineering & Technology, Noida 201303, Uttar Pradesh, India

H. K. D. Sarma et al. (eds.), *Trends in Communication, Cloud, and Big Data*, Lecture Notes in Networks and Systems 99, https://doi.org/10.1007/978-981-15-1624-5_8

few seconds [2]. Detection, diagnosis, and intervention are the major vortices of a clinical practice as the cardiologists need to detect the CVD anomalies, diagnose the disease, and subsequently prescribe the appropriate intervention to alleviate the clinical prognosis of the patients.

According to the latest World Health Organization Fact Sheet report on CVDs published on May 17, 2017, CVDs account for approximately 17.9 million deaths in 2016, contributing to about one-third of all global deaths. Eighty-five percent of the CVD associated mortality is due to heart attack and stroke [1]. Alarmingly, 75% deaths happened in LMICs (low- and middle-income countries) [1]. Death under the age of 70 is termed as premature death, and the WHO data states that while 17 million deaths in 2015 occurred due to non-communicable diseases (NCDs), 82% of these NCD-related mortalities were in LMICs, out of which 37% were caused by CVDs [1]. WHO data suggests that control of tobacco consumption and unhealthy diet along with increased physical activity and curbing the consumption of alcohol will significantly reduce the clinical burden of CVDs. To reduce the burden of morbidity and mortality associated with patients afflicted with CVDs or those with a high risk for an imminent cardiac event, early detection with appropriate clinical intervention and medical counseling are the most effective strategies [1].

Currently, there are different methods available to record ECG like Holter ECGs in which electrical activity of the heart is recorded for several hours and then analyzed, whereas in 12-lead ECGs, the cardiac response is recorded using 12 different leads over multiple heartbeats. Although manual interpretation of ECG traces is in vogue currently, for the ECG interpretation of large volume of data, it is indeed a tedious and time-consuming process. To resolve this issue, we believe that there is a strong need for the implementation of automated technologies to extract the information from ECG traces at a much more faster pace along with a higher degree of precision [3].

Current studies focus on patient classification to diagnose specific diseases based on the overall behavior and analog signal distributions of the ECG traces [4]. On the other hand, new challenges have arisen with the development of wearable devices for the diagnosis of a CVD ailment in a real-time frame. The emergence of big data analytics as a frontier area of artificial intelligence has brought about new challenges like slow speed, lack of storage space for the operation of more advanced computational technologies like machine learning equipped with all the necessary technical requirements for efficient processing and classification of ECG data [5, 6].

Although AI-enabled interpretation of ECG data is a promising strategy, it is wrought with ethical issues as well as data protection issues which envisage additional problems, such as lack of databases available for the verification and validation of its results.

The advent of AI-enabled 3D computer simulations could be prospected as a powerful tool to solve the above-mentioned issues. For the interpretation of ECG signals, it may allow ECG-dependent simulations [7] for facilitating the generation of synthetic data for training and validation purposes.

Long short-term memory (LSTM) network has been used as to classify various types of ECG beats such as premature ventricular contractions (PVCs), atrial premature contraction, paced beats and ventricular couplet from the MIT-BIH database. The significant advancement of LSTM network is that it processes ECG signal without preprocessing it as an input and automatically processes its key features as an output. This study was one of the pioneer studies focused upon the real-time classification based on key features such as the LSTM network [8].

Leutheuser et al. [9], while comparing various techniques enabling the detection of arrhythmia from ECG signals on mobile-based Android application, evaluated from MIT-BIH databases, categorically pointed out the need for accessible, affordable, and reliable computational platforms with abilities to process large volumes of data (big data).

Although multiple tools and techniques have been used to analyze ECG signals, ECG signal processing is still an emerging field due to the lack of awareness of the clinical community about its implementation in terms of accuracy, data security, and robustness. Despite all the frailties of the artificial intelligence-enabled computational platforms, there has been a significant rise in the use of such platforms for enabling clinical decision-making process [10].

One of the complexities associated with machine learning methods that needs to be resolved on a priority basis, pertains to its speed and memory when deployed on portable devices with multiple platforms. In a recent study, support vector machine (SVM) algorithm was integrated into a portable device to classify normal beats, atrial fibrillation, and myocardial ischemia in real time with a sensitivity of 95.1% and specificity of 95.9% [11].

2 Methods

The following subsections give a concise account of the various stages involved in processing of ECG traces along with its image acquisition, processing, and eventual result classification.

2.1 Tools and Technology

In this study, we have used 'R' version 3.4.4 as programming language which is an open-source software equipped with highly advanced image analysis tools and techniques. MySQL version 14.14 Distrib 5.5.60 is an open-source relational database management system (RDBMS) for the creation of an image repository for further processing with LINUX version 14.04 as base platform.

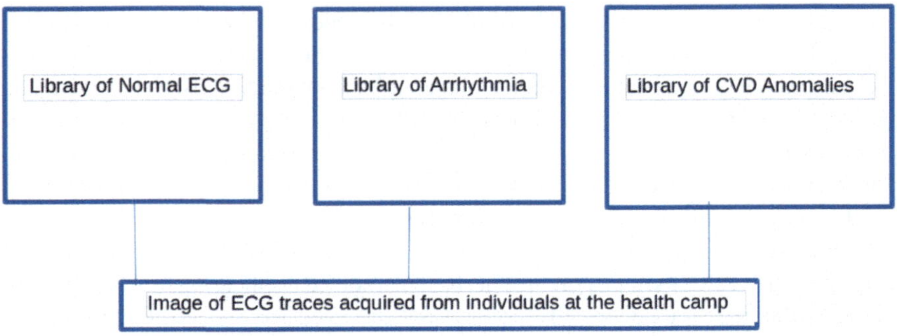

Fig. 1 Classification of ECG traces into different categories: EMD is used for similarity detection among ECGs to desginate appropriate healthy/disease labels

2.2 Data Collection

ECG traces of 315 participants cum patients as primary data were acquired from Batra Hospital & Medical Research Centre, New Delhi, during a cardiovascular health camp organized in Sangam Vihar, New Delhi.

ECG of individuals was recorded during the health camp for diagnosis and further treatment/consultation. Later, the ECG data of all the patients ranging from normal to severe cardiovascular conditions was annotated and categorized by the cardiologists before being processed to create a library of ECG traces capable of being categorized in an automated fashion into normal sinus rhythms (NSRs) or CVD anomalies (Fig. 1).

2.3 Earth Mover's Distance (EMD) as a Base to Classify ECG Traces

EMD has been used extensively for comparing two different images pertaining to content-based image retrieval. The basic concept of EMD relies on the fact that there must be a minimal cost to transform a distribution to a targeted distribution pattern. In the current study, we are using computational implementation of EMD on R-based open-source platforms. The combination that we have proposed could lead to the development of a robust artificial intelligence-enabled tool for faster and precision-oriented classification of ECG traces [12]. EMD(pX, qX) between distributions pX and qX is:

$$\mathrm{EMD}(P, Q) = \frac{\sum_{i=1}^{m} \sum_{j=1}^{n} f_{i,j} d_{i,j}}{\sum_{i=1}^{m} \sum_{j=1}^{n} f_{i,j}}$$

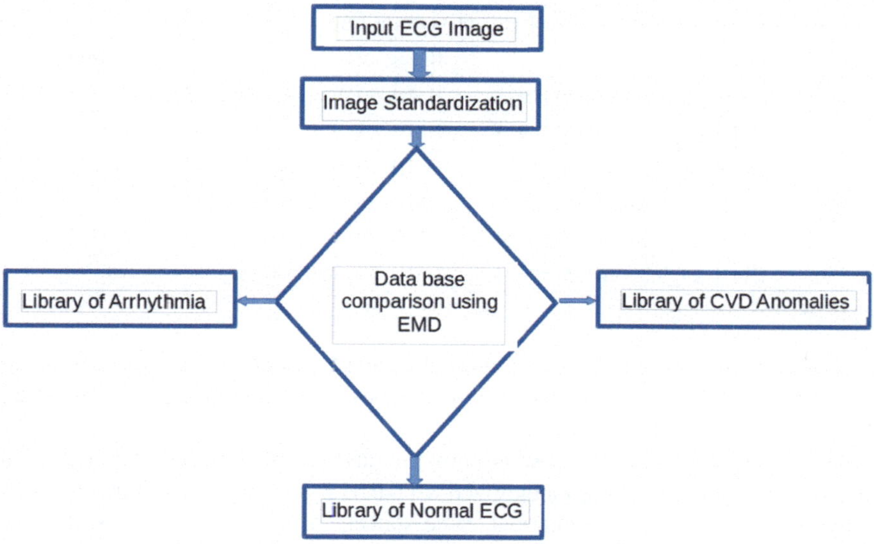

Fig. 2 Flowchart of algorithmic implementation for ECG classification: Screening, processing and differentiating ECG traces

In order to find d_{ij}, one needs to define some distance measure I, and to find f_{ij}, one needs to solve the transportation linear programming.

A comparison between EMD and histogram-based analytics on the basis of robustness revealed EMD to be more robust than histogram matching techniques, because of its inherent ability to overcome the shortcomings of histogram matching techniques with respect to binning of signals/images as well as their quantification on a large scale along with its capability to operate on variable-length representation of the recorded distributions.

Briefly, the data flow diagram (DFD) describes the steps being followed in the current study for the preprocessing algorithm used to screen, process, and differentiate the images using different libraries of ECG traces along with the automated binning of these images into different categories based upon the pixel-based variations in the whole ECG trace (Fig. 2).

3 Results

The salient results from our study suggest that the proposed application would not only enhance the accuracy in classifying the poor-quality ECG traces but also ensure implementation of a computational platform to analyze large volumes of data, a hallmark feature of all health analytics-based studies. Large-scale use of this application would significantly alleviate the mortality rates at primary health-care centers in resource-limited settings found in LMICs. Right now, the proposed application is

Table 1 EMD computation for two selected ECG traces from a given collection

S. no.	Image 1	Image 2	EMD
1	HMH_08_V20180415111923	HMH_29_V20180415112524	4.257542
2	HMH_08_V20180415111923	HMH_08_V20180415111923	0
3	HMH_08_V20180415111923	HMH_36_V20180415135322	4.921893
4	HMH_36_V20180415135322	HMH_36_V20180415135322	0
5	HMH_29_V20180415112524	HMH_36_V20180415135322	4.719507
6	HMH_29_V20180415112524	HMH_29_V20180415112524	0

in validation and testing phase where the different types of ECG images are being recorded to develop a community-specific database from Batra Hospital & Medical Research Centre.

Table 1 depicts the comparisons between random ECG traces obtained from anonymized patient with pre-categorized MIT-BIH repository of ECG traces. Such analyses would give us a numerical value starting from 0 to *n*, where 0 denotes the exact match and concludes the output with the matched library of image like Arrhythmia, Bradycardia, Tachycardia, or any other cardiac anomaly. In case, the EMD result is not 0 even after comparing and matching all ECG digital images pre-stored in the library data sets, and it suggests that there is no exact match. The application then finds the nearest EMD value to 0 and correlates it with the result specifying the image category from the library. If the EMD value is not zero, the searching and sorting mechanism finds the nearest EMD value greater than zero and indicate the ECG category as per the EMD value obtained from the analysis. Thus, the result obtained is fast, accurate, and robust, with capabilities of being scaled up to identify and categorize the cardiac anomalies based upon ECG traces.

The preliminary set of data produced through EMD analysis of the clinically acquired ECG images is depicted in (Figs. 3, 4, and 5).

After comparing the ECG images among each other and numerically evaluating EMD values, we can automatically find specific similarities and differences and can use them to classify ECG traces.

4 Discussion, Conclusion, and Future Scope

Most people suffering from cardiovascular diseases (CVDs) do not have the required health information because of limited access to the healthcare provisioning. The hospital-based registries need to be augmented by community-based data to accurately predict the disease burden. The complexities associated with the landscape of disease burden are indeed having a statutory impact on the delivery of healthcare to the marginalized sections of the society in resource-limited clinical settings predominantly found in LMICs [13].

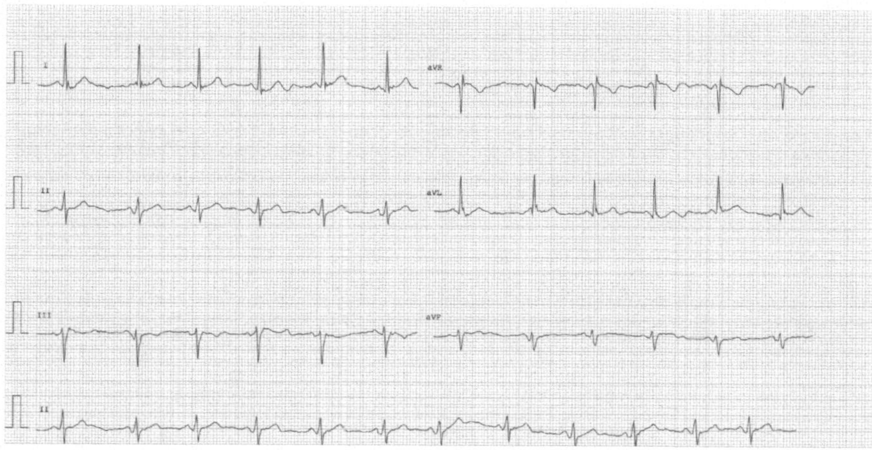

Fig. 3 Sample ECG trace (A)

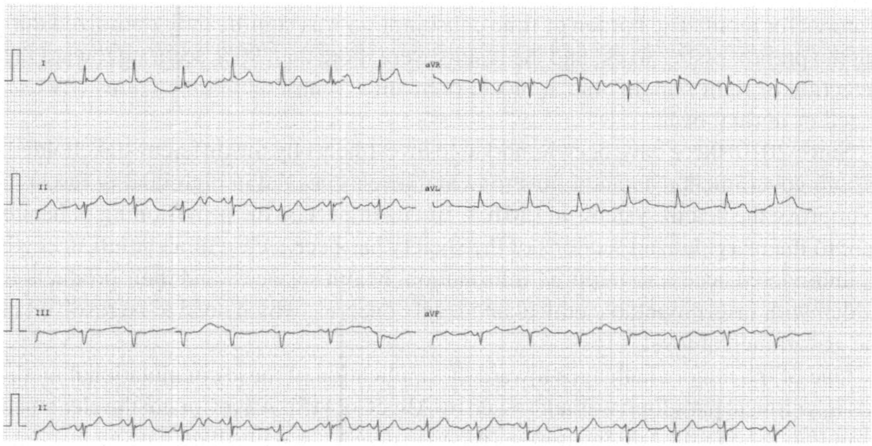

Fig. 4 Sample ECG trace (B)

The incremental progress toward the early diagnosis of CVD has been wrought with challenges. In the past two decades, although numerous technology-driven interfaces have been developed to ensure ease and precision of diagnosis, treatment, and care of the patient, yet the required progress in this field has been slow due to various reasons and circumstances.

Recent advances in technology equipped with computer vision and image analysis are changing the world at an exponential pace. Oresko et al. [14] were one of the pioneer groups to deploy an Android-based wearable platform for real-time ECG acquisition and classification. The development process of ECG classifier was based on a feed-forward MLP neural network. They used the original QRS morphological

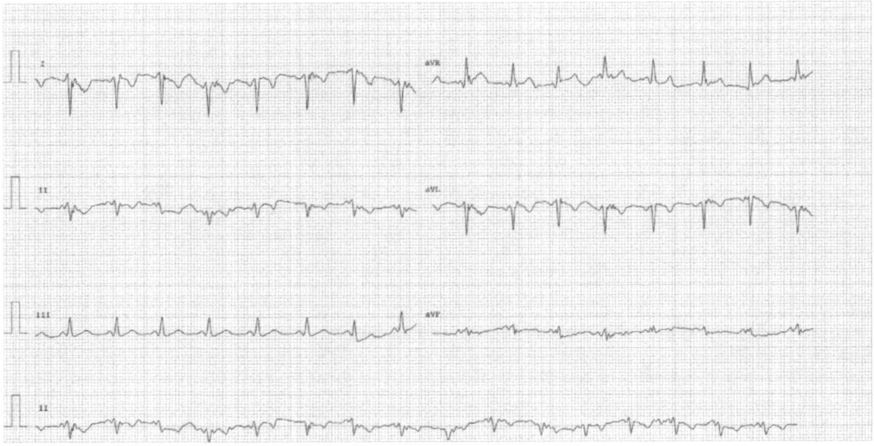

Fig. 5 Sample ECG trace (C)

features for input. The burden of time consumption was significantly optimized during the preprocessing work, and the results were more than 90% accurate toward the detection of right bundle branch block, premature ventricular contractions (PVCs), paced or normal beats.

Since 2010, there have been remarkable investments by the different stakeholders at both governmental and non-governmental levels engaged in high-quality research to improve the health systems in LMICs. But its impact on the provisioning of health care to the marginalized sections of the society has been a case of an uphill struggle as unhindered access to required information, health services, and their products is ridden with insurmountable ethical and legal concerns, which need to be resolved at the governmental level.

One of the major challenges toward the application of these computational methods in a clinical paradigm was addressed by AliveCor [15]. AliveCor got the clearance from Food and Drug Administration (FDA) to detect atrial fibrillation using machine learning algorithm which was integrated into an Android-based application to record ECG and blood pressure (BP) from patients. This unique advancement to predict the ECG data analysis using computational techniques on a real-time frame has augmented its use in various clinics to monitor the ECG traces. A future direction of this research is aimed at unraveling the hidden physiological signals from ECG trace/data in collaboration with the Mayo Clinic [16].

Ethical issues such as patient's rights, safety, and confidentiality have to be the premise for its deployment in large-scale studies to rapidly diagnose CVD anomalies in resource-limited settings of LMICs. Unlike the prior technology-driven revolutions, the impact of the artificial Intelligence-enabled technology might lead to the development of a novel platform hastening the provisioning of health care to large populations in LMICs.

Computational methods have been in vogue to reduce the uncertainty in terms of pseudo-diagnosis for the estimation of cardiac conduction properties from electromagnetic recordings [17]. Similarly, probabilistic models have been introduced to ensure the personalization and integration of uncertainty on data and parameters in cardiac electrophysiology [18].

Abnormalities in human atrial electrophysiology and their consequences on ECG patterns have also been investigated building upon substantial work on multi-scale modeling of atrial dynamics [19]. As in the case of ventricular models, these simple models are not derived from realistic imaging data but are based on a series of assumptions. However, despite being very useful because of their simplicity, these models do not harness the anatomical aspects of the cardiac physiology that would be obtained with imaging data [20–22].

The development of novel computational techniques has led to better understanding of medical problems in terms of ECG signal analysis. Machine learning techniques provide accurate and automatic classifications of heartbeats to detect arrhythmia or cardiovascular anomalies. They have also been used to automate disease diagnosis and analysis of ECG data for which manual inspections are tedious and time consuming. Their adaptability to real-time requirements as well as their embedding on wearable devices has ensured an efficient and reliable monitoring of the ECG activity in hospital settings or at home [13].

Community, age, gender, physical activity, BMI, history of CVD play major roles in the ECG variation of individuals and that is why we need to focus and incorporate every single parameter considered by the cardiologists while interpreting ECG traces. Binning of ECG image repository into its all possible conditions with the help of our proposed algorithm can significantly help in accurate and precision-oriented interpretation and diagnosis of ECG even in remote areas with limited resources to ensure precision-oriented diagnosis, thereby alleviating the clinical prognosis of the patients especially in LMICs, where the scarcity of specialized cardiologists significantly affects the clinical prognosis of patients. Our software is cost-effective and time-sensitive. These features will help in the efficient interpretation of ECGs of individuals, thereby validating its application as a diagnostic tool.

Moving forward, the proposed application will be developed in all three forms: Web-based application, desktop application, and mobile application for facilitating the ease of access as well as its application in the remote areas.

A future area of research pertains to the integration of artificial intelligence-enabled computational platforms not only enabling the ECG diagnosis but also facilitating the provision of closed-loop clinical resource allocation to ensure timely, efficient, and precision-oriented analysis of ECG recordings to diagnose cardiovascular anomalies on a real-time scale.

Acknowledgements This work is supported by a fellowship grant from Indian Council of Medical Research, New Delhi.

References

1. World Health Organization Facts Sheet on Cardiovascular diseases (CVDs) published on 17 May 2017. Last accessed on 12 Nov 2018 http://www.who.int/en/news-room/fact-sheets/detail/cardiovascular-diseases-(cvds)
2. Gurve D, Srivastava AK, Mukhopadhyay K, Prasad NE, Shukla S, Muthurajan H (2016) Electrocardiogram (ECG) image processing and extraction of numerical information. Int J Eng Technol Sci Res. IJETSR 3(7). ISSN 2394–3386
3. Obermeyer Z, Emanuel EJ (2016) Predicting the future-big data, machine learning, and clinical medicine. N Engl J Med 375:1216–1219. https://doi.org/10.1056/NEJMp1606181
4. Rahman QA, Tereshchenko LG, Kongkatong M, Abraham T, Abraham MR, Shatkay H (2015) Utilizing ECG-based heartbeat classification for hypertrophic cardiomyopathy identification. IEEE Trans Nanobiosci 14:1. https://doi.org/10.1109/TNB.2015.2407291
5. Iranyaz S, Ince T, Gabbouj M (2016) Real-time patient-specific ECG classification by 1-convolutional neural networks. IEEE Trans Biomed Eng 63:664–675. https://doi.org/10.1109/tbme.2015.2468589
6. Zhang Q, Chen X, Fang Z, Xia S (2016) False arrhythmia alarm reduction in the intensive care unit using data fusion and machine learning. In: 2016 IEEE-EMBS international conference on biomedical and health informatics (BHI), February, Las Vegas, NV. Piscataway, NJ: IEEE, pp 232–235, 24–27 (https://doi.org/10.1109/bhi.2016.7455877)
7. Cardone-Noott L, Bueno-Orovio A, Minchole A, Zemzemi N, Rodriguez B (2016) Human ventricular activation sequence and the simulation of the electrocardiographic QRS complex and its variability in healthy and intraventricular block conditions. Europace 18(4):iv4–iv15
8. Chauhan S, Vig L (2015) Anomaly detection in ECG time signals via deep long short-term memory networks. In: 2015 IEEE international conference on data science and advanced analytics (DSAA), 19–21 October. Paris, France. Piscataway, NJ: IEEE, pp 1–7. (https://doi.org/10.1109/dsaa.2015.7344872
9. Leutheuser H et al (2014) Comparison of real-time classification systems for arrhythmia detection on Android-based mobile devices. In: Conference proceedings of IEEE engineering medical biology society, pp 2690–2693
10. Ambale-Venkatesh B et al (2017) Cardiovascular event prediction by machine learning: the multiethnic study of atherosclerosis. Circ Res 121:1092–1101. https://doi.org/10.1161/CIRCRESAHA.117.311312
11. Jeon T, Kim B, Jeon M, Lee B-G (2014) Implementation of a portable device for real-time ECG signal analysis. Biomed Eng Online 13:160. https://doi.org/10.1186/1475-925X-13-160
12. Rubner Y, Tomasi C, Guibas LJ (2000) The earth mover's distance as a metric for image retrieval. Int J Comput Vis 40(2):99–121
13. Lyon A, Mincholé A, Martínez JP, Laguna P, Rodriguez B (2018) Computational techniques for ECG analysis and interpretation in light of their contribution to medical advances. J R Soc Interface 15(138):20170821
14. Oresko JJ, Jin Z, Cheng J, Huang S, Sun Y, Duschl H, Cheng AC (2010) A wearable smartphone-based platform for real-time cardiovascular disease detection via electrocardiogram processing. IEEE Trans Inf Technol Biomed 14:734–740
15. AliveCor's heart monitor for iPhone receives FDA clearance. See https://www.alivecor.com/press/press_release/alivecors-heart-monitor-for-iphone-receives-fda-clearance/. Accessed 4 Dec 2018
16. PR Newswire Association LLC. AliveCor and Mayo Clinic announce collaboration to identify hidden health signals in humans. See https://www.prnewswire.com/news-releases/alivecor-and-mayo-clinic-announce-collaboration-to-identify-hidden-health-signals-in-humans-300349847.html. Accessed 12 Dec 2018
17. Wallman M, Smith NP, Rodriguez B (2014) Computational methods to reduce uncertainty in the estimation of cardiac conduction properties from electroanatomical recordings. Med Image Anal 18:228–240. https://doi.org/10.1016/j.media.2013.10.006

18. Konukoglu E et al (2011) Efficient probabilistic model personalization integrating uncertainty on data and parameters: application to eikonal-diffusion models in cardiac electrophysiology. Prog Biophys Mol Biol 107:134–146. https://doi.org/10.1016/j.pbiomolbio.2011.07.002
19. Dokos S, Cloherty SL, Lovell NH (2007) Computational model of atrial electrical activation and propagation. In: Conference proceedings IEEE engineering medical biology society 2007, pp 908–911
20. Adeniran I, MacIver DH, Garratt CJ, Ye J, Hancox JC, Zhang H (2015) Effects of persistent atrial fibrillation induced electrical remodeling on atrial electromechanics—insights from a 3D model of the human atria. PLoS ONE 10:e0142397. https://doi.org/10.1371/journal.pone.0142397
21. Krueger MW et al (2013) Personalization of atrial anatomy and electrophysiology as a basis for clinical modeling of radio-frequency ablation of atrial fibrillation. IEEE Trans Med Imaging 32:73–84. https://doi.org/10.1109/TMI.2012.2201948
22. Seemann G, Höper C, Sachse FB, Dössel O, Holden AV, Zhang H (2006) Heterogeneous three-dimensional anatomical and electrophysiological model of human atria. Phil Trans R Soc A 364:1465–1481. https://doi.org/10.1098/rsta.2006.1781

Pneumonia Identification in Chest X-Ray Images Using EMD

Archit Khatri, Rishabh Jain, Hariom Vashista, Nityam Mittal, Priya Ranjan and Rajiv Janardhanan

Abstract Pneumonia is a common lung infection in which an individual's alveoli fill up with fluid and form a cloudy-like structure. Pneumonia is of two types: (a) bacterial and (b) viral, but both the X-rays have a very similar pattern. The accurate identification along with how much extent the person is infected is still a challenge for doctors. In this paper, the use of EMD to correctly identify infected pneumonia lungs from normal non-infected lungs is shown. EMD, also known as Earth Mover's Distance is the distance of two probability distributions over some region D. First, we preprocessed the images to just have the images of lungs, and then we did some re-scaling, rotation, and normalization of intensity so that we will have a set of uniform size/shape of lungs X-rays, and then, we calculated EMD and compared the results.

Keywords EMD · Pneumonia · X-ray

1 Introduction

Pneumonia is one of the most common lung diseases found in both children and adults. According to recent studies based on pneumonia cases, it is found that all over the world we have around 500 million cases of pneumonia approx., and out of these, around 1.5 million people die each year [1]. The correct diagnostics of pneumonia still remains a problem even for doctors with many years of experience.

A. Khatri (✉) · R. Jain · H. Vashista · N. Mittal · P. Ranjan
Amity Institute of Public Health, Amity University, Noida, Uttar Pradesh 201303, India
e-mail: architkhatri12@gmail.com

R. Jain
e-mail: rishabhjain6416@gmail.com

H. Vashista
e-mail: hariomvashista85@gmail.com

N. Mittal
e-mail: nityam500@gmail.com

R. Janardhanan
Amity School of Public Health, Noida, Uttar Pradesh 201303, India

H. K. D. Sarma et al. (eds.), *Trends in Communication, Cloud, and Big Data*, Lecture Notes in Networks and Systems 99, https://doi.org/10.1007/978-981-15-1624-5_9

Even if the doctors identify the pneumonia in the lungs, they cannot tell about what extent the person has infection because the treatment of the disease depends on it. So, to improve the correct classification of pneumonia detection along with the extent of disease, we designed an algorithm using EMD [2] to differentiate normal lungs from infected pneumonia lungs. This research will be beneficial for doctors to identify the pneumonia with great accuracy. Pneumonia starts when a person inhales germ. It causes swelling in the air sacs in the human lungs, also known as alveoli. The alveoli fill up with the liquid, and it makes difficult for a patient to breathe. This disease can affect anyone, but it can be deadlier for infants, older adults, people with a weak immune system, and people suffering from other diseases like cancer, heart diseases, or diabetes. Some common symptoms of pneumonia may include:

1. Sweating, fever, and shaking chills.
2. Fatigue.
3. Shortness of breath.
4. Vomiting, nausea, or diarrhea.
5. Chest pain when breathing or coughing.

These essential preventions should be followed to stay safe from pneumonia:

1. Make sure children between the ages of two and five years get vaccinated as suggested by doctors. Most doctors recommend flu shots for the child older than six months.
2. Do not smoke as it can cause severe damage to the lungs.
3. Keep your immune system strong by taking enough sleep, eat a healthy diet, and exercise regularly.
4. Always practice good hygiene.

To improve efficiency and accuracy of the diagnosis system, developing an EMD system capable of showing how much extent pneumonia is present in the lungs by taking the X-ray image of the chest as input.

The rest of this paper is ordered as follows:

In Sect. 2, Literature Review is mentioned. In Sect. 3, a brief description of EMD is given with all the mathematical representations. In Sect. 4, the use of EMD is given in the identification of pneumonia. In Sect. 5, all the implementations of EMD in X-ray images are given, and in the next section, Results and Concluding remarks are given. Future Prospect is addressed in Sect. 7 followed by References.

2 Literature Review

Several researchers implemented different models for detecting pneumonia in the lungs using different methods of deep learning and image processing. Sharma et al. [3] made a research on the model of detection of pneumonia cloud using chest X-rays using a very famous approach named image processing. In this research, three lay persons find the ways to automatic diagnostics using medical images. In 2017,

Wan et al. [4] proposed the method of pneumonia risk using online system. In 2016, Ronald et al. [5] researched together in detecting pneumonia using digital images by the method of analyzing ultrasound images. In 2017, Antin et al. [6] researched on a project which detects pneumonia in chest X-rays using supervised learning. In 2017, Ley and Ley-Zaporozhan [7] researched together on topic "Chest CT Imaging in children" and studied the various types of airway diseases in children. Kermany et al. [8] made a research involving deep learning which helped in identifying medical diagnoses results and diseases. Gang et al. [9] researched together in analysis and detection of lung cancer using deep learning techniques in chest X-ray images. Gordienko et al. [10] researched on the model in which they analyze the lung cancer using the chest X-rays by using the dimensionality reduction method in deep learning. They made the analysis on the possible high-quality X-ray images of size 2048 * 2048 pixels. Mao et al. [11] proposed the method of for diagnosing the disease named thoracic using the Generative Classifiers. Ellington et al. [12] researched on the tool for diagnosing the pneumonia with the settings having low resource. Filist et al. [13] researched on the hybrid intelligent model; the task of this model is to detect the segments (analyzed by two-dimensional Fourier spectra) which are related to the diagnosable diseases using the chest X-ray images. Iorio et al. [14] worked on serving the purpose of evaluating the differences between the technologies which are commonly used in diagnosing the pneumonia, the two techniques were lung ultrasonography (LUS) and chest radiography (CR) images, and the results were interesting. Vajda et al. [15] researched on the Automatic Tuberculosis Screening in the chest radiographs, and they used feature selection and other types of analyzing such as culture of sputum smear analysis accuracy of the result which was quite compelling.

As we can see there is a vast scope in development in the detection of lung-based diseases like pneumonia, hence using the Earth Mover's Distance (EMD) could provide better results for the detection of pneumonia.

3 EMD

Earth Mover's Distance (EMD) is a method to calculate the disparity between two multi-dimensional distribution in some space where a distance magnitude between single ones (ground distance) is given. Suppose the two distributions are there, one can be considered as the area with the mass of earth, and the other as a collection of holes in that same area. Then, the EMD is the measure of the least amount of work required to fill the holes with earth. Here the unit of work is the force needed in transporting unit earth by a unit of ground distance. So, it can also be defined as the minimum cost that must be provided to convert one histogram into other. Measuring of EMD is based on a solution of *transportation problem* [16]. For finding mathematical representation, firstly we formalized it as the following linear programming problem:

Let X be the first signature with n clusters, x_i is the cluster representative, and w_{xi} is the weight of cluster.
Let Y be the second signature with m clusters, y_i is the cluster representative, and w_{yi} is the weight of cluster.
Let D be the ground distance matrix, d_{ij} is the ground distance between clusters x_i and y_j.

Let F be the flow matrix, and f_{ij} is the between x_i and y_j. Then,

$$X = \{(x_1, w_{x1}), (x_2, w_{x2}), (x_3, w_{x3}), \ldots, (x_n, w_{xn})\}$$
$$Y = \{(y_1, w_{y1}), (y_2, w_{y2}), (y_3, w_{y3}), \ldots, (y_m, w_{ym})\}$$
$$D = [d_{ij}]$$
$$F = [f_{ij}]$$

Now the WORK $(X, Y, F) = \sum_{i=1}^{n} \sum_{j=1}^{m} f_{ij}\, d_{ij}$
Subject to constraints:

$$f_{ij} \geq 0 \text{ where } 0 \leq i \leq n, 0 \leq j \leq m \tag{1}$$

$$\sum_{j=1}^{m} f_{ij} \leq w_{xi} \quad \text{where } 0 \leq i \leq n \tag{2}$$

$$\sum_{i=1}^{n} f_{ij} \leq w_{yj} \quad \text{where } 0 \leq j \leq m \tag{3}$$

$$\sum_{i=1}^{n} \sum_{j=1}^{m} f_{ij} = \min \sum_{i=1}^{n} w_{xi} \times \sum_{j=1}^{n} w_{yj} \tag{4}$$

The (1) constraint enables mass moving from X to Y. Equations (2) and (3) restrict the amount of mass that can be sent by the clusters in X to their weights and the clusters in Y to receive no more mass than their weights. Equation (4) one forces to move the maximum amount of mass possible. It is also known as the total flow. Once we solve the transportation problem, we will get the optimal flow F. Now the Earth Mover's Distance is defined as the work normalized by the total flow:

$$\mathbf{EMD(X, Y)} = \sum_{i=1}^{n} \sum_{j=1}^{m} f_{ij}\, d_{ij} \div \sum_{i=1}^{n} \sum_{j=1}^{m} f_{ij}$$

4 Use of EMD in Pneumonia Detection

EMD is used in comparing two grayscale images that may differ due to blurring, images containing noise, or local distortions. EMD is also widely used in computing the distances between two histograms of two digital images. In pneumonia detection, it is used in finding the variation of X-ray images of lungs having pneumonia with non-pneumonia ones.

Existing models like pneumonia detection using deep learning have drawbacks like that if any wrong kind of data is feeded into the Deep Net, it can produce very high costs that can hamper the weights of the Deep Net, and it may produce incorrect results. But using EMD, the chances of getting variation in result are very less. Earth Mover's Distance with a threshold ground distance computes very fast as compared to other models.

In the shown images, the first image is a case of non-pneumonia chest X-ray (Fig. 1), and the second is pneumonia-affected chest X-ray (Fig. 2). In the second image, we can see that there is opacity which is mainly due to the presence of white fluids due to presence of bacteria,

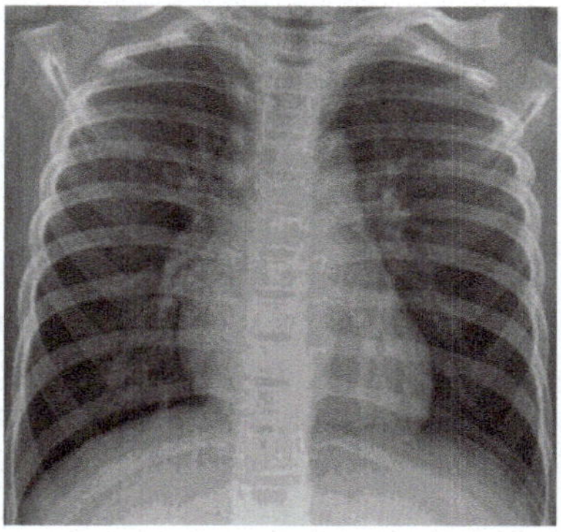

Fig. 1 Non-pneumonia lungs

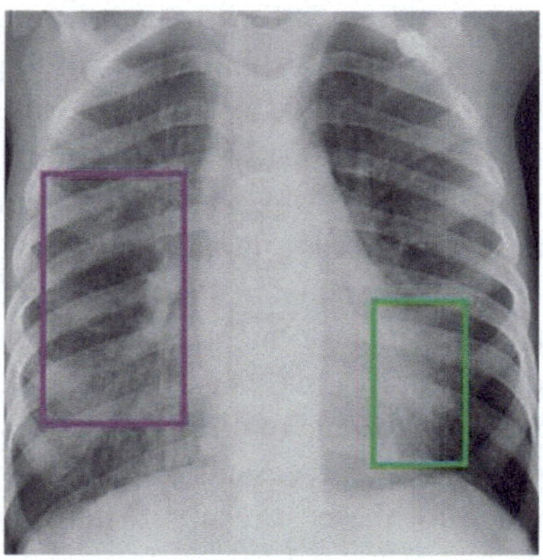

Fig. 2 Pneumonia lungs

which can be seen in the green and purple boxes as compared to the first figure. We used EMD to detect these opaque differences to get desired results.

5 Implementation of EMD in X-Ray Images

To classify images of pneumonia and non-pneumonia X-ray images, doctors usually look for cloudy areas in the lower-bottom part of lungs. The extent of pneumonia depends on how much opacity is there due to presence of these clouds. So, we have used EMD to check the difference between two X-ray images. There is an obvious EMD difference between a pneumonia X-ray image and a non-pneumonia X-ray image, and the EMD difference between two non-pneumonia images is expected to be very less. As EMD is sensitive of little rotations, scaling, and intensity variations, it is must to have images of exact angle, right lungs size, and normalized intensities, else those parameters add error to the results. So, we first preprocessed our images to have same size 500 * 500 pixels per image, we cropped images to only have lungs that spread throughout the image. As it is also done in [17], the X-ray images that are slightly rotated are aligned to have lungs straight symmetrically. Also the intensities of some X-ray images which are too bright or too low are normalized by using the formula (Fig. 3):

$$I_N = (I - \mathrm{I}_{\min}) * (I_{n\max} - I_{n\min})/(\mathrm{I}_{\max} - \mathrm{I}_{\min}) + I_{n\min} \tag{5}$$

where

$I_{\min}$ is minimum pixel intensity of original image
$I_{\max}$ is maximum pixel intensity of original image
$I_{n\min}$ is minimum pixel intensity that we want for normalized image which is 0
$I_{n\max}$ is maximum pixel intensity that we want for normalized image which is 255
I is the matrix of pixel intensity of image.

From [18] taking an example, for an image with minimum pixel intensity value 40 and maximum 190, we want the range of intensities from 0 to 255, we will subtract minimum value from each pixel to get the range from 0 to 150; then, multiply the new range by 255/150 to get the intensity range from 0 to 255 as we have desired.

But in some X-rays, the minimum intensity value is found to be exact 0, whereas a good-quality X-ray photo has minimum value in between 30 and 40. So, this causes huge differences after normalization. So, we also need to make minimum pixel value in the image equal to 40 to even out the X-ray images and then do normalization and then again make the pixels having values less than 20 rounded off to 20 for the best use. By doing this, we normalized the image and also retained the minimum intensity value as average of all X-ray images. Normalization [19] is the most important part of image preprocessing as we want all the pixels intensity to lie in a very specific range for the best results. After image preprocessing, we can successfully use the new images for EMD difference check. Now, first task will be to prepare a set of

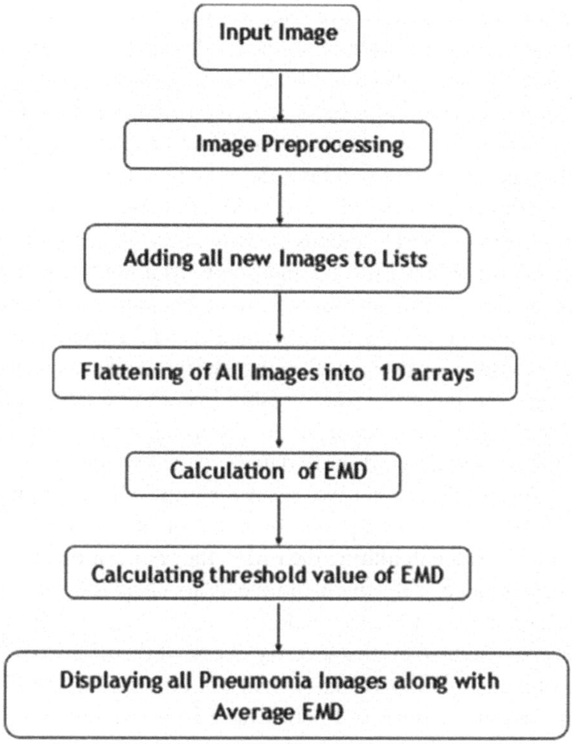

Fig. 3 Flow diagram of detection process

preclassified images into two separate lists with X-ray of pneumonia-infected cases and non-pneumonia-infected cases. We first run the EMD difference check between images of non-pneumonia cases to see the difference between the new set of processed images which should be very less, and the average value is found to be 11. On the other hand, the pneumonia cases have varying extent of disease so EMD variation must be higher than non-pneumonia cases; therefore, we have the average value found to be 26 which is higher than non-Pneumonia cases.

To correctly classify the pneumonia from non-pneumonia X-ray, we ran EMD difference for a single pneumonia X-ray image against all non-pneumonia X-ray images to get the best result. The bigger the set of images, more will be accuracy but more will be processing time. We can also do this for half the images and still can achieve good results. The pneumonia cases have varying cloud opacities so the higher the effect of disease, higher will be the EMD difference between the pneumonia case and set of non-pneumonia images. We found the EMD difference of all the pneumonia X-ray images and put them in a histogram and found a correlation between extent of disease and EMD difference of that image with set of non-pneumonia images. It is now possible to find a threshold value to classify the unlabeled X-ray image to check whether the patient has pneumonia or not [20]. To do that, we used the existing results of EMD of pneumonia X-ray images against non-pneumonia images and used the minimum EMD variation that we found in all the pneumonia images and subtracted the standard deviation of all the EMD results from the minimum value. The standard deviation was subtracted from the minimum value because it is better to have high recall and have more false-negative results than having more false positive because it is better to be cautious even if the disease is not there rather than claiming the absence of disease and later the patient suffers.

After calculating the threshold value, the resulting implementation is ready to be used for real X-ray images classification. We can tune the accuracy by changing the threshold values little up and down. Making the threshold up may increase correct predictions a little up to some extent, but then it will start making incorrect predictions and start classifying all the images as a non-pneumonia person. whereas turning the threshold value down will start classifying all the images as if they had the pneumonia disease in them. So, the threshold should be kept in between of the two extremes. The default value according to the training set is found to be 16 which can be tuned little up and down if more data is feeded into it.

To test the model, we used a sample of 36 preprocessed X-ray images with a mixture of both non-pneumonia and pneumonia images in equal ratios for accuracy check. After calculating the EMD, the results are put in a confusion matrix to evaluate performance. The main motive is to keep the false positive to minimum even if possible, it should also be done to keep false negative low but it has less importance.

We can also do the severity check of the disease from the EMD results as for the worst case; EMD variation will not cross value 80. So, the range of the severity of the disease is from 16 to 80, where 16 is the threshold value. The implementation of the whole system was done in Python 3.6 using Jupyter Notebooks with the use of some helping libraries like cv2 (OpenCV), Matplotlib, Pandas, and scipy.stats.

6 Results

When the model was tested against unlabeled X-ray images, we found that there were only two false-positive cases and six false-negative cases; the rest of the labels were correctly predicted. The average EMD among all the pneumonia images was found to be 26. To get best results, it is must to have correctly preprocessed the images, and they should be well cropped to the borders of the lungs only. After that the only parameter that affects the results is the threshold value which we found by subtracting the standard deviation from fifth percentile of the EMD values of the pneumonia lungs against non-pneumonia lungs. The confusion matrix, the histogram of EMD variation of pneumonia lungs, the most severe case of pneumonia, and the two false-positive results are shown below.

In the confusion matrices in Figs. 4, 5, and 6, it can be observed that decreasing the threshold value starts yielding more false negatives but false positives remain less. On ideal threshold, we have minimum false positives as well as better accuracy. When the threshold value is further increased, it starts yielding more false positives which is red signal to stop.

Fig. 4 Confusion matrix at threshold 15

```
                   Actual True | Actual False
                  |--------------------------
Predicted True    |    16      |     5
Predicted False   |    2       |     13

Accuracy: 0.8056
Threshold value: 15
```

Fig. 5 Confusion matrix at threshold 16

```
                   Actual True | Actual False
                  |--------------------------
Predicted True    |    16      |     4
Predicted False   |    2       |     14

Accuracy: 0.8333
Threshold value: 16
```

Fig. 6 Confusion matrix at threshold 17

```
                   Actual True | Actual False
                  |--------------------------
Predicted True    |    15      |     3
Predicted False   |    3       |     15

Accuracy: 0.8333
Threshold value: 17
```

In Fig. 7 histogram, it can be seen that majority of pneumonia cases are between EMD range 16–25. The severity of disease in this range is low, and these are the results where the use of EMD performs better than other existing classification models. Some

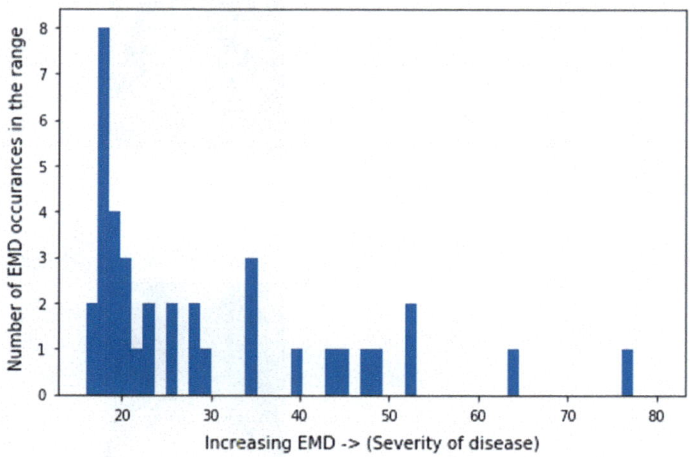

Fig. 7 Histogram of EMD values found on test set (X-axis denotes increasing EMD which corresponds to severity of disease)

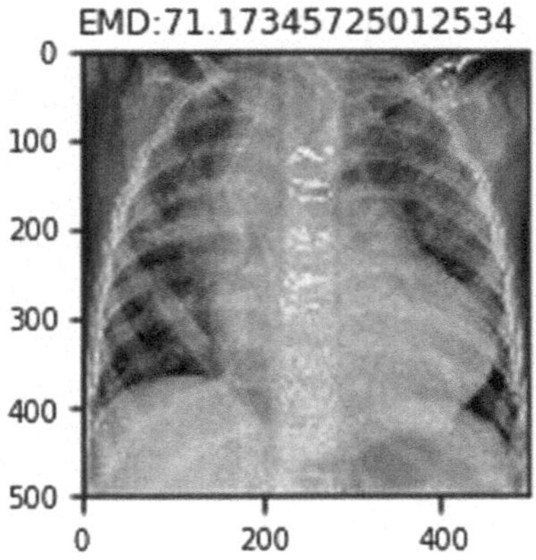

Fig. 8 Most severe case of pneumonia detected by EMD

cases have very high EMD value which means the situation of the disease is quite critical and the patient needs immediate action (Fig. 8).

The most severe case has cloudy-like fluid full in most parts of the lungs. This patient's condition is critical, and immediate action should be advised.

The two false-positive cases were missed by this algorithm to classify as pneumonia-diseased lungs. The reason for that is these X-ray images have very less pneumonia symptoms, and second reason is less light intensity capture in the X-ray. These cases are rare to occur in real life as bad-quality X-rays are discarded and only good-quality X-rays with proper light is used (Figs. 9 and 10).

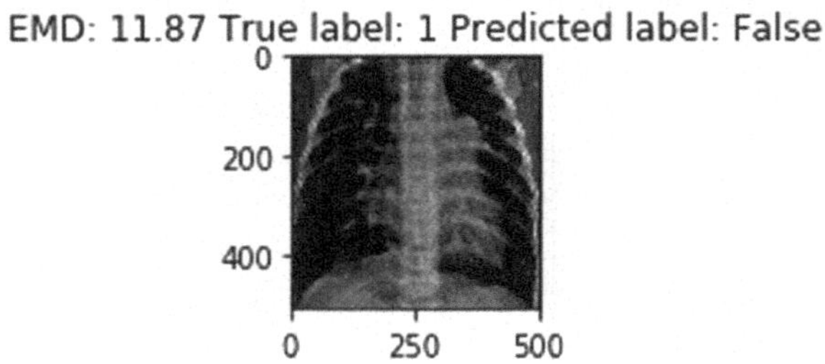

Fig. 9 First false-positive case of pneumonia

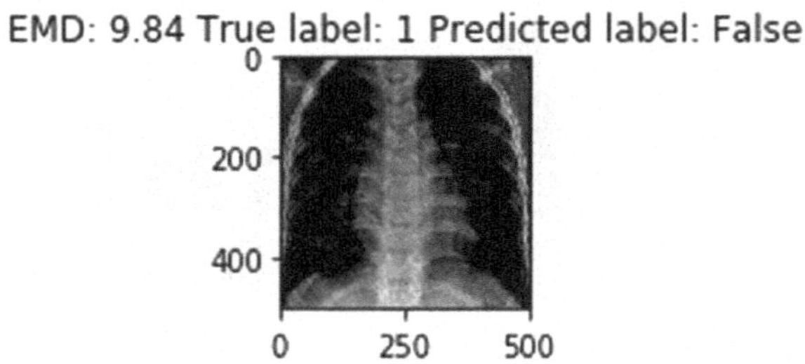

Fig. 10 Second false-positive case of pneumonia

7 Future Prospects

This research will help the doctors to detect pneumonia with high efficiency and accuracy as the system is computerized, and human-prone errors will be minimized because there will be no significant involvement of man in the detection process. It will reduce the cost of detection and gives a substantial change in the field of medical and sciences. This system can undertake further changes if proposed in the future. This research will act as a base for other setups working in the similar field [21]. The few things on which we will be focusing in the future is to make further fine tuning in the developed system to increase accuracy in the results, and the second thing on which light will be thrown is use of colored X-ray images in pneumonia detection as colored X-rays will give us more data in RGB channels, so it will be a huge advantage for EMD classification algorithm to give marvelous results on pneumonia identification in the X-ray.

References

1. WHO (2016) Global pneumonia report 2016. World Health Organization
2. Rubner Y, Tomasi C (2001) The earth movers distance perceptual metrics for image database navigation. In: The Springer international series in engineering and computer science (Robotics: Vision, Manipulation and Sensors)
3. Sharma A, Raju D, Ranjan S (2017) Detection of pneumonia clouds in chest X-ray using image processing approach. In: 2017 Nirma University international conference on engineering (NUiCONE), IEEE, Ahmedabad, India
4. Afifi WNIWM, Warsito IF, Sayahkarajy M, Supriyanto E (2017) The development of an online pneumonia risk prediction system. In: 2017 international conference on robotics, automation and sciences (ICORAS)
5. Barrientos R, Roman-Gonzalez A, Barrientos F (2016) Automatic detection of pneumonia analyzing ultrasound digital images. In: 2016 IEEE 36th central american and panama convention (CONCAPAN XXXVI)
6. Orlova DY, Zimmerman N, Meehan S, Meehan C, Waters J (2016) Earth movers distance (EMD): a true metric for comparing biomarker expression levels in cell populations. In: Allen IC (ed) PLOS One. Virginia Tech University, USA
7. Ley S, Ley-Zaporozhan J (2017) Chest CT imaging in children. Springer International Publishing AG
8. Kermany DS, Goldbaum M, Cai W, Lewis MA, Xia H, Zhang K (2018) Identifying medical diagnoses and treatable diseases by image-based deep learning. DS Kermany – 2018
9. Gang P, Zhen W, Zeng W (2018) Dimensionality reduction in deep learning for chest X-ray analysis of lung cancer. In: 2018 tenth international conference on advanced computational intelligence (ICACI), Xiamen, China
10. Gordienko Y, Kochura Y, Alienin O, Rokovyi O, Peng G, Jiang H, Wei Z (2018) Deep learning with lung segmentation and bone shadow exclusion techniques for chest X-ray analysis of lung cancer. In: Advances in computer science for engineering and education. ICCSEEA 2018. Advances in intelligent systems and computing, vol 754. Springer, Cham
11. Gordienko Y, Kochura Y, Alienin O, Rokovyi O, Stirenko S, Gang P, Hui, Zeng W (2018) Dimensionality reduction in deep learning for chest X-ray analysis of lung cancer. In: 2018 tenth international conference on advanced computational intelligence (ICACI), pp 878–883

12. Ellington LE, Gilman RH, Chavez MA, Pervaiz F, Marin-Concha J, Compen Chang P, Riedel S, Rodriguez SJ, Gaydos C, Hardick J, Tielsch JM, Steinhoff M, Benson J, May EA, Figueroa-Quintanilla D, Checkley W (2017) Lung ultrasound as a diagnostic tool for radiographically-confirmed pneumonia in low resource settings. Respir Med
13. Filist SA, Tomakova RA, Degtyarev SV, Rybochkin AF (2018) Hybrid intelligent models for chest X-ray image segmentation. Biomed Eng 51(5)
14. Iorio G, Capasso M, Prisco S, De Luca G, Mancusi C, Lagan B, Piscopo MA, Comune V (2018) Lung ultrasound findings undetectable by chest radiography in children with community-acquired pneumonia ultrasound. Med Biol 44(8):1687–1693
15. Vajda S, Karargyris A, Jaeger S, Santosh KC, Candemir S, Xue Z, Antani S, Thoma G (2018) Feature selection for automatic tuberculosis screening in frontal chest radiographs. Springer Science + Business Media, LLC, part of Springer Nature 2018
16. Zenteno O, Castaneda B, Lavarello R (2016) Spectral-based pneumonia detection tool using ultrasound data from pediatric populations. In: 2016 38th annual international conference of the IEEE engineering in medicine and biology society (EMBC), IEEE, Orlando, FL, USA
17. Barrientos F, Roman-Gonzalez A, Barrientos R, Solis L (2016) Filtering of the skin portion on lung ultrasound digital images to facilitate automatic diagnostics of pneumonia. In: 2016 IEEE 36th Central American and panama convention (CONCAPAN XXXVI)
18. Chakraborty S, Roy M, Hore S (2017) A study on different edge detection techniques in digital image processing. International Publisher of Information Science and Technology Research
19. Gordienko Y, Gang P, Hui J, Zeng W, Kochura Y, Alienin O, Rokovyi O, Stirenko S (2017) Deep learning with lung segmentation and bone shadow exclusion techniques for chest X-ray analysis of lung cancer. In: The first international conference on computer science, engineering and education applications (ICCSEEA2018)
20. Pradhan A, Sekhar KR, Swain G (2018) Digital image steganography using LSB substitution, PVD, and EMD, Math Prob Eng 2018(Article ID 1804953):11
21. Gianoncelli A, Rizzardi C, Salomon D, Canzonieri V, Pascolo L (2017) Nano-imaging of environmental dust in human lung tissue by soft and hard X-ray fluorescence microscopy. In: Colloquium spectroscopicum internationale XL (CSI 2017), Pisa, Italy

Contemporary Linear Stochastic Models for Forecasting IoT Time Series Data

Rupanka Bhuyan

Abstract In the present times, IoT devices have become ubiquitous for collecting huge volumes of data in real time. These data are of immense value for conducting various types of analyses, critical predictions, and for deriving previously undiscovered insights. Occasionally, because of factors beyond control or errors in the IoT device, anomalies or discrepancies get introduced in the data generated. Therefore, the entire amount of data generated by IoT devices is not always usable; only a part of it is usable. Concepts related to the pre-processing of incomplete data are elaborated in this paper. Also, a number of contemporary stochastic techniques which can be used for forecasting of IoT time series data are discussed in this paper.

Keywords IoT · Pre-processing · Time series · AR · MA

1 Introduction

1.1 Internet of Things (IoT)

The Internet of Things (IoT) which is a very commonly used term nowadays had been coined during the recent past. IoT denotes digital devices or objects equipped with digital functions which can communicate through the Internet [1]. Each IoT device is uniquely identifiable and connectable via radio frequency identification technology (RFID). At present, there are billions of such IoT devices connected across the world providing people to seamlessly communicate with anyone, anything and at anyplace [2].

R. Bhuyan (✉)
ICFAI University Nagaland, Dimapur, India
e-mail: rupanka@yahoo.com

H. K. D. Sarma et al. (eds.), *Trends in Communication, Cloud, and Big Data*, Lecture Notes in Networks and Systems 99, https://doi.org/10.1007/978-981-15-1624-5_10

1.2 IoT Data and Its Importance

IoT devices have become ubiquitous nowadays and are seen almost anywhere and everywhere; as weather sensors, GPS locators, wearable health devices, smart security cameras, Industry 4.0, and so on. These devices operate in real time and generate data at high velocities [3]. The amount of data generated also grows in volume at an exponential rate very rapidly. Some IoT devices collect or generate data for a single parameter (say, room temperature) while others do so for multiple parameters (for instance, heartbeat, blood pressure, and walking speed). This has become possible due to the easy availability of the devices at reasonably low cost. In this paper, we refer to the data generated or collected by any IoT device or sensor as IoT data.

One observation is that most of the data generated by IoT devices are temporal in nature, i.e., there is a time element in the data. This makes IoT data very critical for temporal analytics, forecasting future events, deducing causes of past events, etc.

1.3 Some Issues with IoT Data

These IoT devices or sensors mostly operate on their own with minimal or no human intervention. Sometimes due to factors beyond control or due to unexpected faults in the concerned IoT device, improper data is collected or not collected at all. Over ruling the fact that, IoT devices are generally made to be robust and fault tolerant, we still encounter cases where faulty data is generated (noise) or the device fails to collect data. Factors beyond control, such as environmental or other unforeseen reasons may be directly or indirectly responsible for this type of errors.

Regardless of the reasons, whatsoever the data generated by the device cannot be used in its entirety. The various types of analysis methods or analytical studies cannot be applied to this type of inappropriate or missing data.

Another important issue mostly pertaining to multivariate IoT data is that the various parameters are heteroscalar, i.e., their scales of magnitude may vary largely. For instance, an IoT sensor in an automobile may record *engine_rpm* in the range {0, 1, 2, 3, ..., 3000} and *gear_position* in the range {0 (reverse), 1, 2, 3, 4, 5}. Although the data is correct, the same cannot be supplied as input to the various algorithms as it will result in erroneous results or predictions. This is due to the difference in units and scale of the data. The parameters with larger scales will have more influence on the results of the underlying prediction or forecasting algorithm.

1.4 Pre-processing the IoT Data

For the reasons highlighted in Sect. 1.3, the data collected from the IoT device needs to be processed to make it usable. A few of the popular techniques used to do this are discussed below.

Handling Missing Data. Missing data can be taken care of mostly by (i) deletion or (ii) imputation.

(i) *Deletion*: In some cases, the records containing missing information are simply discarded. This results in a dataset containing complete information. However, it reduces the original size of the dataset [4]. This is detrimental for some of the modern forecasting techniques which are data-intensive and uses artificial intelligence, machine learning, etc.
(ii) *Imputation*: Sometimes generating or calculating the missing data is more important than simply deleting the incomplete information. Some of the commonly used techniques for doing so are *mean imputation*, *last value carried forward*, *imputation based on logical rules*, *random deterministic imputation*, *random regression imputation*, etc. [4]

Scaling Data. In case of multivariate data it is usually seen that the minimum and maximum values of the variables vary widely, as discussed in Sect. 1.3. This has an effect on the forecasting or clustering/classification algorithm to be used. For instance, the variables with wider ranges will have greater influence on the results while the variables with relatively smaller ranges will have little or no effect on the same. This will produce biased or incorrect outcomes.

To eliminate this problem all the variables have to be scaled down to the same levels of magnitude. A number of methods to achieve this have been proposed, such as *standardization* and *normalization*.

Standardized values for a variable x can be generated as

$$x_{\text{standardized}} = \frac{x - M}{S} \tag{1}$$

where M is the mean value of x and S denotes the standard deviation [5].

Consider a variable x having minimum and maximum values x_m and x_M, respectively. Then the new normalized value of x is given by

$$x_{\text{normalized}} = \frac{x - x_m}{x_M - x_m} \tag{2}$$

The model (1) and (2) are detailed in [5].

2 Time Series

2.1 Introduction to Time Series

We can define a time series as a sequential or chronologically ordered set of data points. Typically, these data points are measured over successive times. In mathematical notation, it can be expressed as a random variable vector x_t [6–8] with t denoting the time dimension ($t \geq 0$).

If multiple variables are expressed in a time series at a time 't' then it is a multivariate time series otherwise it is univariate. The values of x_t can be either continuous (e.g., continuous air pressure) or discrete (e.g., population of a city recorded annually).

2.2 Characteristics of a Time Series

There are four characteristics in a time series, namely trend, cyclicality, seasonality, and irregularity.

A time series may tend to go upward, go downward, or remain stagnant over time. If it goes upward as in the price of gold, it exhibits upward trend; likewise, downward trend is exhibited in a time series depicting mortality rates.

If variations based on the season (such as increase in sales of cold drinks during summer) are exhibited by a time series then it has seasonality.

Repetition of certain phases in a time series in cycles is known as the cyclicality of the time series.

Variations in the time series (which are mostly extreme in nature) caused by events beyond control such as natural disasters, wars, etc., are classified as random variations or irregularities.

3 Stochastic Forecasting

3.1 Introduction

In stochastic time series modeling, we assume that a time series $\{x_t, t \geq 0\}$ follows a probability model describing the joint distribution of the random variable x_t.

The probability structure of a time series can be described by a mathematical expression, i.e., a stochastic process [7]. And, depending on whether the current value of the series is a linear or nonlinear function of past observations, a stochastic time series model can be linear or nonlinear. In this paper, only the linear models are discussed.

3.2 Linear Models

In terms of understandability and implementation, linear models are relatively simpler than the nonlinear ones. The most widely used linear models which can be used for forecasting IoT time series data are discussed in this section.

Autoregressive Model (AR): The autoregressive model, namely AR(m) [7, 9, 10] is where the future value 'f_t' of the variable to be forecasted is calculated as a linear combination of a determined number of past observations, say 'm'. Added with these observations are (i) a random error (say, 'E_t'), and (ii) a constant (say, 'K'). The general form of the model [7] can be denoted as:

$$f_t = \sum_{i=1}^{m} f_{t-i} T_i + E_t + K \tag{3}$$

Here, 't' represents a time instant. The model uses some parameters represented by the terms 'T_i' ($1 \leq \mathrm{i} \leq m$). An AR model which considers the past 'm' number of observations for forecasting is categorized as a model of order 'm'.

Yule-Walker equations [7] can be utilized to estimate optimal values for 'T_i'.

Moving Average (MA): Contrary to the AR model, the past errors are used for forecasting the future values in the MA(n) model [6, 7, 10] furnished in (4).

$$f_t = \sum_{j=1}^{n} E_{t-j} P_j + E_t + M \tag{4}$$

where 'n' denotes the order of the MA model, 'M' expresses the mean of the series. Also, the parameters of the model are expressed as 'P_j' ($1 \leq \mathrm{j} \leq n$). The random errors are considered as a white noise process, i.e., a sequence of independent and identically distributed random variables with a mean value of zero and a variance σ^2 which is constant [6, 7]. Details of model (3) and (4) are available in [7].

Autoregressive Moving Average (ARMA): A more generalized model can be derived from the models listed in (3) and (4) to result in something called the ARMA (m, n) model as shown in (5).

$$f_t = \sum_{i=1}^{m} f_{t-i} T_i + \sum_{i=1}^{q} \theta_j \varepsilon_{t-j} + E_t + K \tag{5}$$

The terms 'm' and 'n' [6, 7, 10] represent the orders of both the models, respectively.

A few operators used to manipulate these types of models are listed below.

$$\boldsymbol{L f_t = f_{t-1}} \tag{6}$$

$$T(L) = 1 - \sum_{i=1}^{m} T_i L^i \quad (7)$$

$$P(L) = 1 + \sum_{j=1}^{n} P_j L_j \quad (8)$$

The models depicted in Eqs. (6), (7), and (8) are contributed by Hipel et al. [7] and consequently Cochrane et al. [6].

Autoregressive Moving Average (ARIMA): Some of the commonly encountered time series are non-stationary; they have trends and are seasonal. This cannot be handled by the ARMA model. The seasonality component of a seasonal time series can be removed by finitely differencing the data points with the ARIMA(m, d, n) approach [6, 7, 9, 11] with 'd' representing the level of differencing. Mathematical expression Eq. (9) represents the model by Hipel et al. [7] and Lombardo et al. [12]:

$$T(L)(1 - L)^d f_t = P(L)E_t \quad (9)$$

The parameters 'm' and 'n' have the same meanings as in the Eqs. (3) and (4).

Seasonal ARIMA: The above-listed models are not adequate for dealing with seasonal time series data. A more generalized model [7, 9, 13] has been proposed by Box and Jenkins to deal with this which is a modification of ARIMA. The model is not included here. The seasonal difference is calculated as $z_t = f_t - f_{t-s}$ as stated by Hyndman [14] where $s = 12$ for monthly time series data and $s = 4$ for a quarterly one.

4 Model Selection

While it may not be very easy to choose an appropriate linear stochastic model to forecast values of a specific IoT data time series, Box and Jenkins [9] have suggested a procedure for selecting an appropriate one.

Although the detailed procedure given by Box and Jenkins is beyond the scope of this paper, the fundamental steps for carrying out the method are furnished below:

Step 1: Assume a general ARIMA model class
Step 2: Identify a tentative model
Step 3: Estimation of the parameters of the model identified in Step 2
Step 4: If the adequacy of the model is 'not acceptable' go to Step 2
Step 5: Select model for forecasting.

5 Conclusions

IoT devices are the present and the future for ubiquitous data generation and collection. These data are generated at extremely high velocities and scale. Due to their huge volume and speed they are quite useful for deriving analyses, critical for real-time forecasting, and discovering unprecedented insights. One inherent quality of IoT data is that there is an implicit temporal order to the collected data.

However, in the ground reality, due to various physical or device-related or environmental factors which are mostly beyond control, the data collected may gain some percentage of errors, noise, etc., which renders them unusable for the various analysis techniques. Some of the essential techniques for processing this collected data have been discussed in this paper.

Further, this processed temporal data can be used for predicting or forecasting future values, trends, etc., which is of immense value for the concerned. In this context, a class of stochastic forecasting models has been observed to produce comparable results with some of the modern models used, such as support vector machines, artificial neural networks. In this paper, linear stochastic forecasting methods which are in practice contemporaneously have been discussed highlighting their inherent and fundamental characteristics.

References

1. Li L (2013) Technology designed to combat fakes in the global supply chain. Bus Horiz 56(2):167–177
2. Perera C, Liu C, Jayawardena S (2015) The emerging internet of things marketplace from an industrial perspective: a survey. IEEE Trans Emerg Top Comput. https://doi.org/10.1109/TETC.2015.2390034
3. Wang P, Valerdi R, Zhou S, Li L (2015) Introduction: advances in IoT research and applications. Inf Syst Front 17:239–241. https://doi.org/10.1007/s10796-015-9549-2
4. Gelman A, Hill J (2006) Missing-data imputation. In: Data analysis using regression and multilevel/hierarchical models (Analytical methods for social research). Cambridge University Press, pp 529–544. https://doi.org/10.1017/cbo9780511790942.031
5. Zheng A, Casari A (2018) Feature engineering for machine learning, 1st edn. O'Reilly Media, Inc. USA, pp 30–31
6. Cochrane John H (1997) Time series for macroeconomics and finance. University of Chicago, Spring Graduate School of Business
7. Hipel KW, McLeod AI (1994) Time series modelling of water resources and environmental systems. Elsevier, Amsterdam
8. Raicharoen T, Lursinsap C, Sanguanbhoki P (2003) Application of critical support vector machine to time series prediction. In: ISCAS '03. Proceedings of the 2003 international symposium on circuits and systems, vol 5, pp V-741–V-744, 25–28 May 2003
9. Box GEP, Jenkins G (1970) Time series analysis: forecasting and control. Holden-Day, San Francisco, CA
10. Lee J, Univariate time series modeling and forecasting (Box-Jenkins method), Econ 413, Lecture 4, https://sites.google.com/site/junsoolee/teaching, Accessed on 18 July 2018
11. Zhang GP (2003) Time series forecasting using a hybrid ARIMA and neural network model. Neurocomputing 50:159–175

12. Lombardo R, Flaherty J (2019) Modelling private new housing starts in Australia. In: Pacific-Rim real estate society conference
13. Hamzacebi C (2008) Improving artificial neural networks' performance in seasonal time series forecasting. Inf Sci 178:4550–4559
14. Hyndman R, Athanasopoulos G (2018) Forecasting. OTexts, Heathmont, Vic.

A Journey from MD5 to SHA-3

Sandip Ghoshal, Pradosh Bandyopadhyay, Surojit Roy and Monalisa Baneree

Abstract If X is sender of a message and Y is the receiver of the same, then X encrypts the same with Y's public key and then sends the encrypted data to Y. It confirms about authenticity and authorization of receiver. If X sends message and Y receives, X encrypts the message with X's private key and sends the encrypted data to Y. It confirms authenticity and authorization of the sender [1]. Such a schema could work properly. It involves the usage of a message digest or hash. Hash is a fingerprint or the summary of the message. It carries similar concept of cyclic redundancy check (CRC). Integrity of the data is verified with this process. This process actually confirms that the data should not be damaged between the path of sender and receiver [2]. Hashing confirms few things like (i) complexity of calculation of hash value of a message should be decreased; (ii) it follows a one-way encrypting procedure or technique; and (iii) different hash values should be generated by a particular hashing technique for any two different messages.

Keywords Encrypt · Public key · Private key · Message digest · Hash · Cyclic redundancy check (CRC) · MD5 · SHA family · RIPEMD-160

1 Idea of a Message Digest or Hashing

Both super-speed and tight security are demanded by the latest cryptographic applications [3]. Basically, in the time of sending messages, they are encrypted by the senders either to confirm message security or to ensure authenticity of the sender.

S. Ghoshal (✉) · P. Bandyopadhyay · S. Roy · M. Baneree
Department of MCA, Techno India, Salt Lake, Kolkata, West Bengal, India
e-mail: ghoshal.sandip@gmail.com

P. Bandyopadhyay
e-mail: pradosh.banerjee@yahoo.com

S. Roy
e-mail: roy.surojit@gmail.com

M. Baneree
e-mail: monalisa_erjee@yahoo.com

H. K. D. Sarma et al. (eds.), *Trends in Communication, Cloud, and Big Data*, Lecture Notes in Networks and Systems 99, https://doi.org/10.1007/978-981-15-1624-5_11

Sender can apply this thing either by using public key of the receiver which confirms message security or by using own private key which confirms authenticity of sender. In this regard, technique of hashing is used to hide the original data from the outside world. This technique also gives us an assurance or confirmation about the user's authentication.

Hashing can be of two types: (i) keyed hashing [hash value can be generated by using a fixed value, known as key. That means ORIGINAL VALUE <OPERATION> KEY VALUE = HASH VALUE] and (ii) non-keyed hashing [key value or any fixed value is not required in this process. That means OPERATION ON ORIGINAL VALUE = HASH VALUE].

In case of a non-keyed hashing, output depends only on original input or message. On the other hand, in case of a keyed hashing, output generates by using both input and key value. In case of cryptographically secure cases, the concept of non-keyed hashing is used, and in addition to generate the used key, brute force provides the best result among all the popular and existing algorithms;

The following example shares the concept of keyed message digest or keyed hash:

Suppose our original input is 6000 and key value is 6. We are performing division as operation. Dividing 6000 by 6 will always result as 1000. So 1000 is the fingerprint or hash value of 6000. If we alter either 6000 or 6, then we will not get result as 1000. Another important thing is only by knowing the key value 6 or hash value 1000, no one can guess or generate the original information which we want to hide from the others. So we can also say that the fingerprint of a message does not advise anything about the original data. This is because there are various other possible equations, which can generate the result 1000.

The following example shares the concept of non-keyed message digest or non-keyed hash:

Now suppose here our original input is 123456. We want to perform non-keyed hashing technique over the data. We have added each digit with its next digit and then collected the last digit from the result to get the hash value. Advantage of this procedure is that the size of the hash value is much lesser than the size of the original input.

The basic requirements of hashing technique:

- For a message or input, the generated hash value of a particular hashing technique should be same always, and the complexity of the calculation should be very low.
- In any moment by using a hash value, the original message or input should not be regenerated.
- For any two different massages, a hashing technique should always generate two different hash values.

If we get the same hash value for two different messages after using the same hashing method, then that is actually violating the principle, which is known as a collision. The message digest algorithms generally create a message digest or fingerprint of length 128 bits or 160 bits. This means that the possibilities of any two hashes being the same are one in 2^{128} or 2^{160}, respectively. Clearly, possibility of collision can happen only in theory but very few in practical. Here we are going to converse about two chief hashing algorithms, MD5 and SHA.

2 Theoretical Background

2.1 MD5

Authentication of user and data integrity can be maintained by hash functions. MD5, SHA-1, and RIPEMD-160 are the most popular hash functions in the current computer age. All these popular algorithms are designed on the basis of the algorithm of MD4 [4]. MD5 is a message digest algorithm which is designed by Ron Rivest. Algorithm of MD5 has developed on the basis of other algorithms of MD family. That means MD5 has few predecessors algorithms. All the algorithms of MD-Series are developed and designed by Ron Rivest. The initial or first message digest algorithm was known as MD. He then released the next version, with name MD2. But according to the requirement, those two algorithms were very much weak and he started working on his next algorithm which was known as MD3. But this time, all his efforts were in vain. That algorithm was collapse, and for that reason, he did not release the same. Then, Rivest designed MD4. However, within a few years, MD4 was also proved to be imperfect. Therefore, Rivest designed and released the upgraded version of MD4 with name MD5. MD5 generates fingerprint very quickly and accurately. It provides 128-bit message digests. Over the years, researchers have developed possible weakness in MD5. So far MD5 creates no collision for any two different messages. But it is not guaranteed in the long run. After few early processing, the input text is converted to 512-bit blocks (they are more alienated into 1632-bit sub-blocks). Finally, the algorithm produces a set of four 32-bit blocks, which builds the 128-bit message digest.

2.2 SHA [Secure Hash Algorithm] Family

The Secure Hash Algorithm (SHA) is developed by the National Institute of Standards and Technology (NIST) with the collaboration of NSA. SHA was published in the year 1993. It was published as a Federal Information Processing Standard (FIPS PUB 180). Next, we got a revised version as FIPS PUB 180-1 in the year 1995, and the name was given as SHA-1. We can also say it was a modified version of MD4. SHA takes input of lesser than 264 bits in length and provides output of 160 bits in length (32 bits extra than the message digest formed by MD5). The word **secure** came from following two merits of SHA:

(1) Original input can be retrieved never and
(2) Any two messages cannot have same message digest.

The **Secure Hash Algorithm** is a collection of cryptographic hash functions designed by the National Institute of Standards and Technology (NIST) as a U.S. Federal Information Processing Standard (FIPS). It includes the following set of algorithms:

SHA-0: In the year 1993, it was published under the name "SHA." Within a short period, it was withdrawn by them, and they also released another version with name SHA-1 which was a modified version of the previous one.

SHA-1: It was capable to provide 160-bit message digests. It was similar to the earlier MD5 algorithm. This was designed by the National Security Agency or NSA. Till 2010, this technology was used in very large scale.

SHA-2: This supports similar hash function with different block sizes. They are recognized as SHA-256 and SHA-512. They vary in their word size; SHA-256 accepts in 32-bit words where SHA-512 uses 64-bit words. SHA-224 is another available version of SHA-256, and SHA-384 is another truncated version of SHA-512. These were also designed by the NSA.

Designed by the U.S. National Security Agency (NSA) published **SHA-2** in the year 2001 by the NIST as a U.S. Federal Information Processing Standard. It is a set of cryptographic hash functions (**SHA-224, SHA-256, SHA-384, SHA-512**). SHA stands for Secure Hash Algorithm. SHA-2 is the combination of the following four hash functions $\rightarrow$ 224, 256, 384, and 512 bits.

In the year 2005, we got this version of SHA when a SHA-1 became slightly insecure due to some mathematical or computational fault. Although SHA-2 carries some similarity to the SHA-1 algorithm, SHA-2 can prevent all those attacks which made SHA-1 insecure.

In the year 2012, we got the newer version of the family, namely SHA-3. It was completely a new algorithm comparing to SHA predecessors.

SHA-3: In the early days of its development, it was known as **Keccak**. It takes the equal hash lengths as SHA-2, and the inside construction of the algorithm differs considerably from the rest of the SHA family. NIST has said that FIPS 180-5 contains SHA-3.

SHA-3 is designed by Guido Bertoni, Joan Daemen, Michaël Peeters, and Gilles Van Assche, building upon RadioGatún.

Keccak got the first prize in the NIST hash function competition on October 2, 2012. SHA-3 is not the replacement of SHA-2. As all MD-5, SHA-0, and SHA-1 received several attacks and to some extent those algorithms failed to prevent few of them, then NIST eagerly wanted to make a different algorithm which is not based on any previous member of SHA family.

The sponge construction is used by SHA-3. In this algorithm, message blocks are XORed into the early bits of the state, which is then invertible permuted. In SHA-3, the result consists of a 5×5 array of 64-bit words, 1600 bits in total. The authors demand 12.5 cycles per byte on an Intel Core2 CPU. However, in hardware implementations it is remarkably quicker than all other finalists.

3 Application Area

3.1 Verifying the Integrity of Files or Messages

A very much vital function of secure hash is verification of message integrity. Checking whether any changes have been done to a message (or a file) or not is the main application of these algorithms. This is simply done by comparing between previously stored message digest and newly accepted message digest.

For the same, nearly all digital signature algorithms only confirm the authenticity of a hashed digest. If fingerprint is authentic, then definitely the original message is also authentic.

MD5 and SHA1 hashes are sometimes check only verification of integrity. This practice is not safe; because of chain of trust problem, we cannot trust on posted hashes. Even if they cannot be trusted as files, except these Web sites are authenticated by HTTPS; but in this case, the hashes are obsolete.

3.2 Password Verification

Password verification is another area of application. It is similar to the previous application. Generally, password is not present or stored in simple or plain text. Instead, it is stored in digest form. When user provides his/her password, then the application first converts the same into digests and checks with the stored digest for authentication. That means no one can retrieve the original password at any moment. If user losses or forgets the password, then that should be replaced by the new one. Generally, password is concatenated with a prefix non-secret value. This value is known as salt value. PBKDF2, Bcrypt, or Scrypt are the key stretching functions. These functions help us to raise the time requisite to execute brute force attacks on password digests which are already stored.

3.3 File or Data Identifier

File or data can be identified by relevant message digests. Each file has some set of attributes through which they can be identified uniquely. For example, file size, source or file location, types of file are the necessary information through which a file can be identified easily and uniquely. Files which are on peer-to-peer file sharing networks can also be identified or recognized by the hash functions or digests. For example, fingerprint of file is combined with corresponding size of file, original location of file, content information in the application of MD4 and ed2k. Another example of the same is magnet links. These kinds of hashes are also known as a hash tree which uses for extra benefits or top hash of a hash list.

4 Conclusion

So, here we can see, the main motive of all these is to implement a program or algorithm which can generate unique message digest or hash for each existing input. Starting from MD5 to moving around SHA-3 is a big journey so far. Though we got a popular, effective algorithm like SHA-3, still we are looking for an algorithm which is effective but simpler to implement and understand.

Acknowledgements We thank to all the experts who have developed and designed the template.

References

1. Kahate A (2013) Cryptography and network security, 3rd edn. McGraw-Hill
2. Stallings W (2017) Cryptography and network security. Pearson
3. Khan E, El-Kharashi MW, Gebali F, Abd-El-Barr M (2007) Design and performance analysis of a unified, reconfigurable HMAC-hash unit. IEEE Trans Circuits Syst I Regul Pap 54(12):2683–2695. https://doi.org/10.1109/tcsi.2007.910539
4. Bettilyon TE (2017) An introduction to hashing in the era of machine learning. Published by the researchers of Google and MIT, Dec 2017

An Approach to Achieve Compression Along with Security via a User Assigned Security Key with Possible Lossless Compression

Ashis Datta and Mousam Saikia

Abstract Although the current version of this research work might not be fool-proof as this kind of works are currently being researched upon and yet to be successfully implemented at large. Popular software such as WinRAR makes use of AES to encrypt the data and also compresses it at the same time but only one algorithm is used to compress the data [1]. It is sometimes seen that using different lossless compression algorithms multiple times can achieve a better compression ratio and may even achieve better security [2]. Our plan is to achieve such kind of compression ratios while achieving confidentiality with a model that we have devised ourselves.

Keywords Data compression · Loss-less · Encryption · Decryption · Cipher · Private key · Symmetric key · Plain text · Cryptographic modeling · Digital right

1 Introduction

1.1 Encryption

The process of converting from plaintext to cipher-text is known as enciphering or encryption. Some classical encryption techniques [3] include symmetric cypher model, substitution techniques, transposition techniques, rotor machines and steganography. Our algorithm partially falls into the category of symmetric cypher model. In this model, the sender encrypts the plain text into some cypher text using some key to encrypt the data. The same key has to be used to decrypt the data. Our algorithm modifies this concept by making sure that the original content that is sent by the sender can be decrypted only in a specific system i.e., of the receiver. We achieve this by incorporating the receiver's system's MAC address into the final key that is used to encrypt data.

A. Datta (✉) · M. Saikia
Sikkim Manipal Institute of Technology, Majitar, Sikkim 737136, India
e-mail: ashis.d@smit.smu.edu.in

H. K. D. Sarma et al. (eds.), *Trends in Communication, Cloud, and Big Data*, Lecture Notes in Networks and Systems 99, https://doi.org/10.1007/978-981-15-1624-5_12

1.2 Compression

There are two basic types of data compression [4, 5]; lossy data compression and lossless [6] data compression.

In lossy data compression techniques, the original data upon decompression will lose or undergo changes in some of its minute features which are not easily noticeable by humans. This type of compression technique is used in compression of images, audio and video data.

Lossless compression techniques are those in which, upon decompression will produce an exact replica of the original data. Our software makes use of this type of algorithms to achieve compression. The technique we have used should be able to compress general-purpose data, i.e. the algorithm will accept any bit string given to it.

We present a new algorithm for lossless data compression and encryption. The basic idea of the algorithm is to define a unique compression and encryption of files specified by a user based on a key/password provided by the user. We are making use of lossless data compression [7] algorithms such as Snappy, Gzip, BZip2 and XZ. The key specified by the user is taken as an input and files given are compressed multiple times depending upon the result of XOR operation between the password and the MAC (Media Access Control) address of the user's system which is automatically retrieved by the software and is unique to each system.

During the process of decryption and decompression of the data, the contents will be available at a temporary sandboxed location in a folder and a process will be created that will keep on running till its closed or the system is turned off. The extracted file(s) will remain secure as copying, distributing or tampering with it will not be possible. This is achieved by running the decryption/decompression software as non-administrator. This part of the software will terminate itself if it is being ran with administrator privileges. The contents will be deleted if the process is killed and the space in the hard drive the content takes up will be filled with random data such that a hacker cannot retrieve that data later. This will be achieved by replacing the content of the data just before deleting them.

1.3 The Problem of Piracy

The way we consume media is changing and as content owners and producers look for new ways to keep audiences engaged, the challenges of protecting content across multiple screens becomes increasingly difficult. Innovation, of course, is a good thing, but as new ways of streaming content are developed, it is actually aiding digital piracy.

Piracy is detrimental to innovation, directly affecting job creation and economic growth. Industries protect their ideas through a variety of legal instruments such as patents, copyrights, designs, models and trademarks. Without the protection of their intellectual property rights [8], they may be less inclined to develop new ideas and products. Risks are particularly high for industries in which the research and development costs are high compared to the production costs of the finalised product. Faced with a diminishing turnover due to counterfeiting and piracy, industry investment in research and innovation could well slow down. This would limit development, growth and competitiveness, forcing industries to simply close or at least limit production.

The problem of piracy is that it's here to stay and it could be argued that the problem is growing; in August it was reported that Game of Thrones was the world's most pirated TV show, with 1.6 million illegal downloads in just four weeks. This accounted for over a quarter of all pirated downloads from the top 100 torrent sites. Another recent report from Viacess-Orca highlighted that during the last Football World Cup they monitored 20 million illegal downloads. What this demonstrates is that the pirates appear to be winning. However the definition of piracy encompasses many different layers: from the amateurish camera recordings shot discreetly from a cinema, through to Internet-streamed content that has been cracked and then uploaded to an app store.

The many layers of piracy certainly means that content owners and producers can never fully protect their content, they can only do as much as possible to limit what pirates can access, and this is where it gets interesting. To implement a DRM (digital rights management) solution for streamed content across all devices is both complicated and expensive, so what we're seeing is that studios are choosing to focus their protections around high value digital media such as High Definition and 4 K over standard definition content. What they need is content protection that follows the content and therefore is device agnostic.

1.4 DRM (Digital Rights Management)

Implementing an effective DRM security is the answer and as most content publishers will argue, it is vital to facilitate continued innovation in digital media. As new devices that stream apps and host media players are constantly being developed, the need for secure and unobtrusive digital distribution is urgent. In order for DRM solutions [9, 10] to work to protect the sale of books, films, and music that is growing online, the critical component of a digital key, which allows a user or device to decode the protected content - is required [11]. However, even the best encryption schemes are useless if a hacker can quickly acquire the key.

The costs of DRM security [12] breaches are significant; with the Motion Picture Association of America (MPAA) estimating it costs the film industry $6 billion per year in lost revenue. The weakness with DRM security and implementation is that it can be easily hacked and cumbersome to users if strong application-level defences are not leveraged. In an attempt to curb piracy, many DRM vendors are resorting to using invasive digital rights protection techniques that assume the system is always under attack by pirates, which causes restrictions and performance degradation to honest users.

Since most streaming media applications are performance intensive, DRM security solutions mustn't noticeably impact performance; hence it's not an option. For a DRM system to be well received, it's imperative that the original content simply looks and works better than the pirated copy. The solution lies in imposing DRM security strategies that are effective at preventing piracy, while not degrading the consumer's experience.

DRM vendors are always going to be held hostage to the speed of technological innovation. The problem here, is that the quality of the very content that we on our different devices is likely to be undermined by the massive loss of revenue caused by piracy.

2 The Scheme

The algorithm has two modules with two main parameters each. The modules are briefly described below:

The file processor: This module is responsible for the generation of a special form of file that is both an encrypted and compressed version of a user assigned file.

The reverser: This module is responsible for producing the exact replica of the original file on acquiring the proper file and key (the file that was processed with the key that was assigned during the execution of the previous module).

Both the modules have one common parameter i.e., the key assigned by the user, which needs to be entered during separate instances. The first instance being the processing of an unprocessed file and the second instance being the decryption and decompression of a processed file. During the processing of an unprocessed file the second parameter will contain the path and name of the unprocessed file in the form of a string. While obtaining the original file, the second parameter will contain the path and name of the previously processed file also in the form of a single string. The programming language used for the development of this project is java.

3 Literature Survey

Serial number	Author name	Paper title	Features
1	Robert Franceschini and Amar Mukherjee	Data compression using encrypted text	• It establishes that encrypted representation of text leads to substantial saving of storage space • It is an interesting dictionary based compression method with better performance than gzip and arithmetic coding

(continued)

(continued)

Serial number	Author name	Paper title	Features
2	Matt Mahoney	Data compression explained	• Covers a lot about lossless data compression algorithms and commonly used lossy data compression algorithms • Most of the lossless data compression algorithms are used in this project
3	Mark Stamp	Digital rights management: the technology behind the hype	• Covers topics such as DRM tethered and untethered systems • Covers media snap DRM in detail
4	Chung-Ping Wu	Design of integrated multimedia compression and encryption systems	• It was shown that security could be achieved without sacrificing the compression performance or the processing speed
5	Cappaert, Nessim Kisserli, Dries Schellekens and Bart Preneel	Self-encrypting code to protect against analysis and tampering	• Covers topics such as code obfuscation and self-modifying code • Shows how two or more files are dependent upon one another and achieves self-modifying code
6	M. VidyaSagar, J.S. Rose Victor	Modified run length encoding scheme for high data compression rate	• Covers the run length encoding scheme in details and how the modified version is better from other algorithms in terms of both compression and performance. But better compression algorithms are available
7	Alan Story	Intellectual property and computer software	• Shows how proprietary software can be prevented from piracy • Shows the various ongoing problems for such software • Shows how to tackle these kind of problems

(continued)

(continued)

Serial number	Author name	Paper title	Features
8	Kun-Won Jang, Chan-Kil Park, Jung-Jae Kim and Moon-Seog Jun	A study on DRM system for on/off line key authentication	• This paper shows an algorithm that can encrypt files by dividing it into blocks. Conventional systems can replay digital content only after entirely decrypting it • It can be used to provide free demos for users
9	Pasi Tyrväinen, Jarmo Järvi, Eetu Luoma	Peer-to-peer marketing for content products combining digital rights management and multilevel marketing	• It states "out of the $32 billion market for music in 2002 only about $0.09 billion is sold by paid downloads and online retailers accounted for a mere 1% of music sales" • Shows a model for distribution of copyrighted content in a peer-to-peer network
10	Mikko Löytynoja, Tapio Seppänen, Nedeljko Cvejic	Experimental DRM architecture using watermarking and PKI	• The paper describes a unique architecture for DRM and discusses watermarking in detail

3.1 Objective of This Paper

The primary objective of the paper is to develop a unique algorithm for compression and encryption of files specified by a user based on a key/password provided by the user. We are making use of lossless data compression algorithms such as Snappy, Gzip, BZip2 and XZ. A key comprising of integers that has a variable length from 1 to 19 digits specified by the user is taken as an input and files given are compressed multiple times depending upon the result of XOR operation between the password and the MAC address of the user's system which is automatically retrieved.

It is intended to be used in industries where piracy causes huge amounts of loss. With more research into this field, copyright laws will benefit a great deal in the future. It also has the possibility to become the norm for downloading and uploading of copyrighted content from the internet.

3.2 Solution Strategy

We propose a solution strategy by applying the following steps:

- By continuously running an algorithm to clear the clipboard as a thread in the background. This thread is stopped when the program is closed.
- Upon exit, the files previously extracted to a temporary location will be forcibly deleted.
- The user won't have write access to the files as modifying or coping the contents of these files will make this project useless.
- The key passed by the user will undergo XOR operation with his/her MAC address which is unique for every computer and the result will be used for the encryption instead of the actual key.

4 An Implementation

4.1 Architecture Diagram

In Fig. 1,

T Represents the original unprocessed file.
C Represents the first module.
k Represents the key.
*T Represents the file after processing.

In Fig. 2,

*T Represents the processed (encrypted and compressed) file.

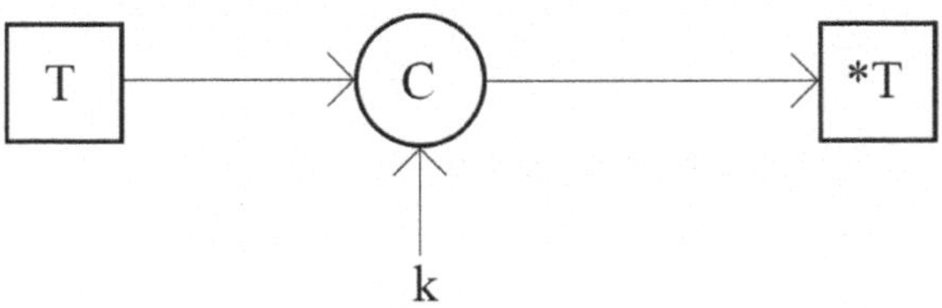

Fig. 1 Block diagram for the first module (compresses and encrypts)

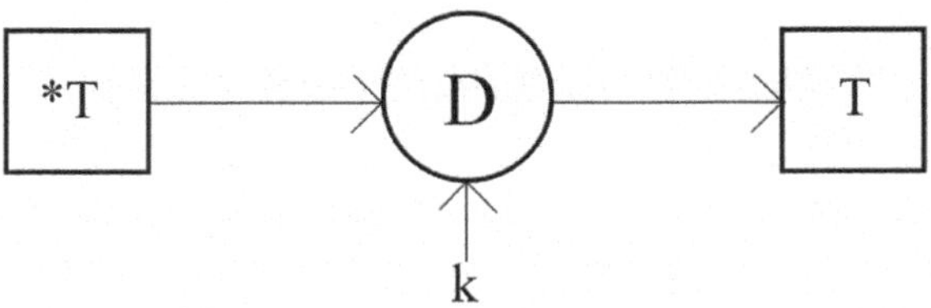

Fig. 2 Block diagram for the second module (decompresses and decrypts)

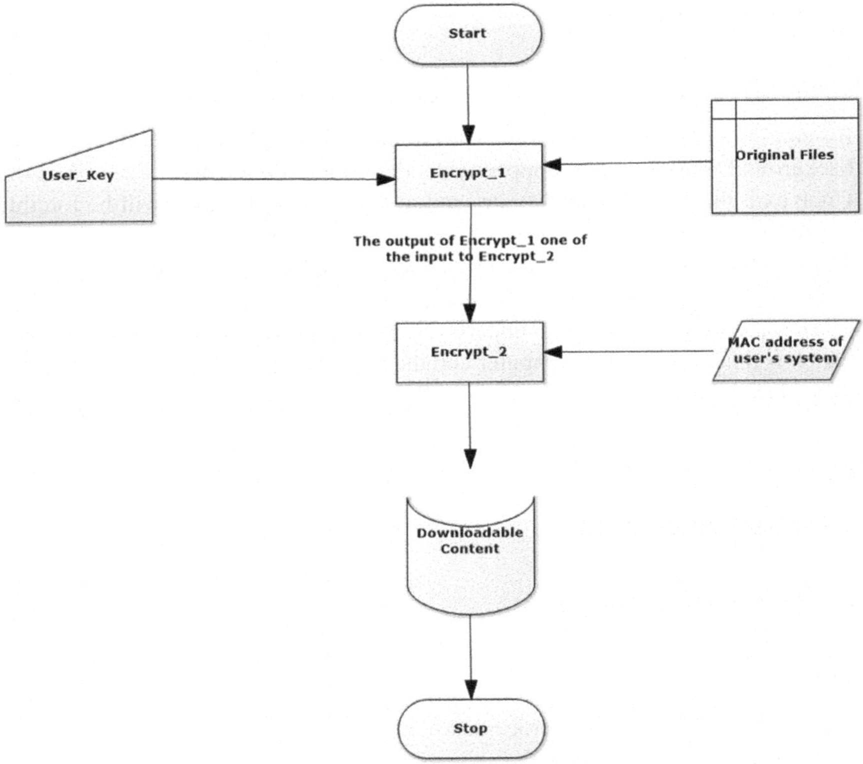

Fig. 3 The file processor/the first module

D Represents the reverser module.
k Represents the key.
T Represents an original copy of the unprocessed file.

4.2 Detailed Diagram

See Fig. 3.

4.3 Detail Process and Pseudo Codes

Compression of data in a lossless manner by making use of various algorithms more than once sometimes achieve better compression ratios. But this kind of algorithms are currently looked into and researched upon without any assurance that it is always possible. However, the algorithms we are using are quite simple and easy to use. The source codes of these algorithms are easily available on the internet. Snappy is a fast compression and decompression technique achieving speeds of 250 MB/s for compression and 500 MB/s for decompression on a single core i7 of 64 bit. It is based on LZ77 algorithm which achieve compression by replacing repeated occurrences of data with references to a single copy of that data existing earlier in the uncompressed data stream. XZ compression algorithm is a dictionary based compression technique. Gzip makes use of one of the algorithms already defined within itself depending upon the type of data it is given to compress. BZip2 makes use of Burrows-Wheeler algorithm to compress data. Our algorithm can compress multiple files but it is not an archival tool. Although Gzip is an archival tool, this algorithm will not make use of the ability because the file(s) after decryption and decompression will be found at a temporary location in a temporary folder. The software will be fairly easy to use. First, the user needs to specify the file(s) to be compressed and then in the second step, a password is asked to be entered by the user and the system's MAC address is retrieved which then undergoes XOR operation with the password provided. One assumption of our algorithm is that the password entered will be of numbers only with a maximum length of nineteen digits and a minimum length of one digit. The file(s) then undergo lossless compression via various algorithms according to the result generated during the XOR operation. BZip 2, Snappy, Gzip, XZ are lossless data compression algorithms [13] used in building this software. Their algorithms are described below. However, in these algorithms, a file is created and the data output stream is written onto it. In our algorithm, these functions instead of creating an entire file, will return the output stream which will be further compressed. The output file will be created only when it has finished compressing that stream. The resulting stream will also be encrypted.

The pseudo code for BZip 2 compression algorithm in java:

Get the original file.
Create an empty output file.
Initialize output stream.
Read content of the original file.
Initialize BZip2 output stream.
Apply Burrows-Wheeler transform on the data and compress.
Write data to the stream.
Write contents of the stream to output file.
Close output stream.

The pseudo code for BZip 2 decompression algorithm in java:

Get compressed file.
Store contents of compressed file in byte array.
Initialize integer variable n to 0.
While (Reading of data in input stream is not complete, set n as counter)
{

Decompress the contents of the input stream.
Write contents to the output buffer.

}
Close all input and output streams.

The pseudo code for Snappy compression algorithm in java:

Enter file to be compressed.
Initialize input stream to be compressed from the input file.
Initialize the output file.
Initialize the output stream.
Initialize encoder.
Apply Snappy algorithm.
Write the output to the output stream.
Create the output file.
Close output stream.

The pseudo code for Snappy decompression algorithm in java:

Read contents of compressed input file.
Put the content in a byte buffer.
Initialize Snappy decoder.
Decompress the contents by setting the correct properties.
Put decompressed content in an output stream.

The pseudo code for Gzip compression algorithm in java:

Initialize input stream.
Initialize the output stream.
While (Reading of data in input stream is not complete)
{

Apply Gzip algorithm to data.
Write to the output stream.

}
Close output stream.

The pseudo code for Gzip decompression algorithm in java:

Get compressed file.
Store contents of compressed file in byte array.
Initialize integer variable read_bytes to 0.
While (Reading of data in input stream is not complete, set read_bytes as counter)
{

Decompress the contents of the input stream.
Write contents to the output buffer.

}
Close all input and output streams.

The pseudo code for XZ compression algorithm in java:

Initialize input stream.
Initialize output stream.
Set size of dictionary.
While (Reading of data in input stream is not complete)
{

    Apply XZ Algorithm.
    Write compressed contents to output stream

}
Close output stream.

The pseudo code for XZ decompression algorithm in java:

Get compressed file.
Store contents of compressed file in byte array.
Initialize integer variable read_bytes to 0.
While (Reading of data in input stream is not complete, set read_bytes as counter)
{

    Decompress the contents of the input stream.
    Write contents to the output buffer.

}
Close all input and output streams.

The code for function to detect whether the program is being run as administrator:

Initialize a Boolean flag.
Initialize a string with the command reg query "HKU\S-1-5-19"
Run the command in command prompt.
If no error is found set the flag to false.
Else set the flag to true.

A pseudo-code for one round of the encryption process

int key_1, key_2, secret_1, pass_value;
if(key_1 is not divisible by secret_1 and key_2 is not divisible by secret_1)

    use BZip 2 on file_1;

if(key_1 is not divisible by secret_1 and key_2 is divisible by secret_1)

    use Snappy on file_1;

if(key_1 is divisible by secret_1 and key_2 is not divisible by secret_1)

```
    use GZip on file_1;
if(key_1 is divisible by secret_1 and key_2 is divisible by secret_1)
    use XZ on file_1;
pass_value = XOR(key_1, key_2);
return (file_1, pass_value);
```

In the pseudo-code described above, key_1 is used as the result of XOR operation between the password provided by the user and the MAC address. The key_2 is a private key that will be present with authenticated users only, it can also be used for recovery purposes; secret_1 is a simple integer value that is kept private; file_1 is a byte stream and pass_value will be used for further encryption.

5 Results and Discussions

The expected outcome of this project is that it shall be used in industries where piracy causes huge amounts of loss. With more research into this field, copyright laws [14] will benefit a great deal in the future. It also has the possibility to become the norm for downloading and uploading of copyrighted content in the internet. Since, this algorithm produces a lossless copy of the original content, high quality multimedia content will consume reasonable amount of hard disk space while being secure (Fig. 4).

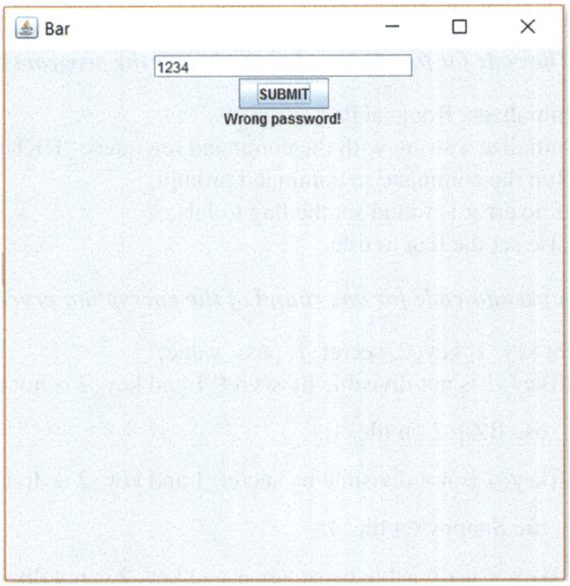

Fig. 4 The form for getting the password (in this case, a wrong password is entered)

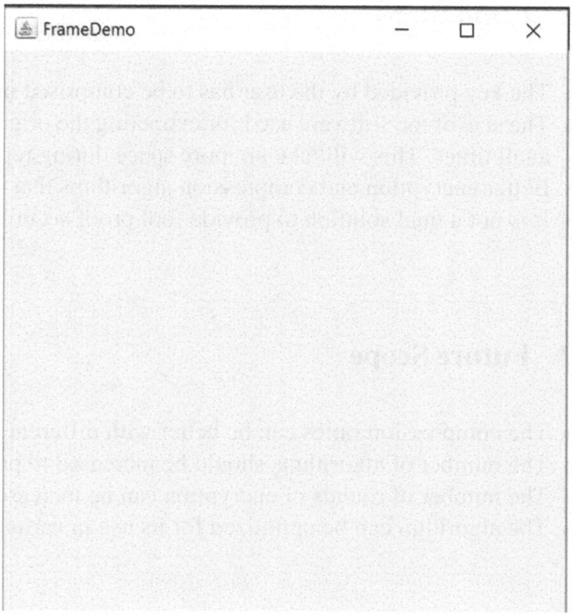

Fig. 5 This window will appear on successful entry of the password

When the Fig. 5 appears, a new directory named temp_folder will appear which will contain the decompressed files. The user will be unable to perform paste operation when this window is showing. Upon closing this window, the files will be deleted and will reappear again when the program is rerun.

Currently, we are able to achieve at best a 55–60% compression ratio for pdf file of 14 Megabytes. Some other tests include images, music and audio-video files, 60–65% in case of images, 80–90% in case of music and audio-video files. The time required for compressing depends upon the file size and the type of file. However, this technology seems to be working best with document files.

Below is a figure of a music file that has been compressed and encrypted successfully (Fig. 6).

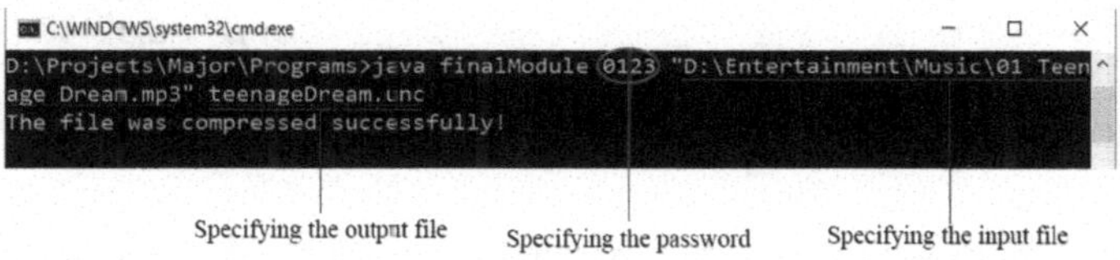

Fig. 6 A file successfully compressed

6 Limitations

- The key provided by the user has to be comprised of integer values only.
- The size of the software used for extracting the original files will remain the same at all times. This will take up more space during replay of the content.
- Better encryption and compression algorithms that work individually exist.
- It is not a final solution to provide fool proof security or data compression.

7 Future Scope

- The compression ratios can be better with different settings.
- The number of algorithms should be increased to provide better security.
- The number of rounds of encryption can be increased for better security.
- The algorithm can be optimized for its use in various regions.

8 Conclusions

This work is expected to help in industries where piracy auses huge amounts of loss. With more research into this field, copyright laws will benefit a great deal in the future. It also has the possibility to become the norm for downloading and uploading of copyrighted content in the internet. Since, this algorithm produces a lossless copy of the original content, high quality multimedia content will consume reasonable amount of hard disk space while being secure.

Acknowledgements Let us express our heartiest gratitude to respective authorities of Sikkim Manipal Institute of Technology, Sikkim, INDIA for providing resources used during the entire development process.

References

1. VidyaSagar M, Rose Victor JS (2013) Modified run length encoding scheme for high data compression rate. Int J Adv Res Comput Eng Technol (IJARCET) 2(12), December 2013
2. Zhang Y, Zhang LY, Zhou J, Liu L, Chen F, He X (2016) A review of compressive sensing in information security field. IEEE Access 4:2507–2519
3. Cappaert J, Kisserli N, Schellekens D, Preneel B (2006) Self-encrypting code to protect against analysis and tampering
4. Franceschini R, Mukherjee A (1996) Data Compression using encrypted text. In: Proceedings of Advanced Digital Libraries (ADL'96)

5. Mahoney M (2013) Data compression explained. Dell Inc.
6. Khelifi F, Brahimi T, Han J, Li X (2018) Secure and privacy-preserving data sharing in the cloud based on lossless image coding. Signal Process 148:91–101, ISSN 0165-1684
7. Alsheikh MA, Lin S, Niyato D, Tan HP (2016) Rate-distortion balanced data compression for wireless sensor networks. IEEE Sens J 16:5072–5083
8. Story A (2004) Intellectual property and computer software May 2004 intellectual property rights and sustainable development
9. Tyrväinen P, Järvi J, Luoma E (2004) Peer-to-peer marketing for content products combining digital rights management and multilevel marketing. In: Proceedings of IADIS International Conference on e-Commerce 2004 (EC2004) Lisbon, Portugal, 14–16 Dec 2004
10. Löytynoja M, Seppänen T, Cvejic N (2003) Experimental DRM architecture using watermarking and PKI. In: Proceeding of First International Mobile IPR Workshop Rights Management of Information Products on the Mobile Internet, Helsinki, Finland, 47–52
11. Jang KW, Park CK, Kim JJ, Jun MS (2006) A study on DRM system for on/off line key authentication. In: Proceedings of the 2006 International Conference on Security and Management, SAM 2006, Las Vegas, Nevada, USA, June 26–29
12. Stamp M (2003) Digital rights management: the technology behind the hype. J Electron Commer Res (August, 2003)
13. Wu CP, Jay Kuo CCJ (2005) Design of integrated multimedia compression and encryption systems. IEEE Trans Multimed 7(5), October 2005
14. De Hert P, Papakonstantinou V, Malgieri G, Beslay L, Sanchez I (2018) The right to data portability in the GDPR: towards user-centric interoperability of digital services. Comput Law Secur Rev, 34(2):193–203, ISSN 0267-3649

Landslide Susceptibility Mapping: Development Towards a Machine Learning-Based Model

Sonam Lhamu Bhutia, Samarjeet Borah and Ratika Pradhan

Abstract Landslides are movement of rocks, debris or earth along a slope. There are seven states into which the activities of a landslide can be classified into; they are dormant, relict, suspended, reactivated, active and stabilized. Landslide susceptibility map is very crucial as it provides significant data and information that is required for planning of landslide-prone areas as it provides crucial information about spatial probability regarding the occurrence of landslide which is very important in the planning of land use. This paper provides a review of various studies and experiments conducted in this domain of research. The focus area of the study is Sikkim Himalayan region as it shares some similar type of parameters.

Keywords Landslide susceptibility map · Machine learning · Causative factors · Accuracy · ROC

1 Introduction

The movement of rock, earth and debris down a slope can be defined as landslide [1]. Landslides are driven by the force of gravity and are result of failure of the materials that makes a hill.

They are the most common natural hazards that occur in the mountain and hilly regions, causing substantial damage to both property and life. Sikkim being a hilly region is highly vulnerable to landslides. Hence, in order to assess the factors that contribute towards the occurrence of landslide in any area it is vital to understand its process. This in turn will help us to predict the future landslides, analyse its hazard

S. L. Bhutia (✉) · S. Borah · R. Pradhan
Sikkim Manipal Institute of Technology, Sikkim Manipal University, Majhitar, Sikkim, India
e-mail: sonamlhamu10@gmail.com

S. Borah
e-mail: samarjeetborah@gmail.com

R. Pradhan
e-mail: cse_ratika@yahoo.co.in

H. K. D. Sarma et al. (eds.), *Trends in Communication, Cloud, and Big Data*, Lecture Notes in Networks and Systems 99, https://doi.org/10.1007/978-981-15-1624-5_13

and prevent the damages caused due to landslides and also device suitable mitigation measures and models [1].

However, it is difficult to project an accurate time frame regarding the occurrence of landslides and thus landslide susceptibility map plays a vital role in providing crucial information in for management of the landslides.

1.1 Landslide Susceptibility Map

Landslide susceptibility map is very crucial as it provides significant data and information that is required for planning of landslide-prone areas as it provides crucial information about spatial probability regarding the occurrence of landslide which is very important in the planning of land use [2]. Main objective of the landslide susceptibility map is to identify the areas where the likelihood of occurrence of landslide is the most. It is defined as the likelihood of a landslide occurring in an area on the basis of local terrain conditions [3].

It is the degree to which a terrain can be affected by slope movements, i.e. an estimate where landslides are likely to occur [4] which involves the spatial distribution and segregation of the terrain units according to their level of vulnerability to landslides and provide the related information related to the probability of any new landslides occurring in an area.

Over the years, though many assessments for developing models for landslide susceptibility mapping have been carried out worldwide, it still stands as one of the major areas of research because different regions would have different economic, political, social and cultural factors and thus the technique of assessment and management for landslide may vary from one region to other.

1.2 Landslide Causes

Instability in slope is usually the cause of the landslide; however, there are various other factors that make the slope susceptible to landslides [1, 3].

The factors that cause the landslide can be categorized as geological, morphological, physical and man-made. While there may be several causes for a landslide, however, there are very few triggering factors, i.e. single event that finally initiates the landslide.

The main triggering factors for the landslide are rainfall, volcano, glacier and earthquake. There may be several factors associated with landslide however; we normally focus on single triggering factors.

2 Related Works

Many assessments for developing models for landslide susceptibility mapping have been carried out worldwide. These involve both qualitative and quantitative techniques, which have been developed over time. However, the quantitative techniques have gained more popularity in the recent decades as it has not only the ability to incorporate more causative factors but also various probabilistic methods [3].

Both the quantitative and qualitative techniques for landslide susceptibility mapping have their own advantages and disadvantages.

Over the years, much research has been done. Accordingly, the approaches for the development of a model can be grouped into mainly statistical, soft computing and analytical [5].

The area of the study may vary accordingly; for a relatively small study area development of model for landslide susceptibility model, the various approaches applied may yield a better accuracy. However, when the area of study is comparatively large, the analytical approach may fail to produce a better accuracy; hence, today most of the researches are focused towards the statistical and soft computing approaches [5].

Sharma et al. [6] have incorporated Shannon's entropy along with the information value-based statistical model in order to study the susceptibility of landslide in the study area, Sang area, East Sikkim.

Here, the thematic layers corresponding to all the causative parameters were considered and were overlapped, and thus, as a result 78,256 numbers of polygons were formed. Then, TLIV, i.e. total landslide information value, was calculated for every polygon. Shannon's entropy value of the individual parameters was then multiplied with the total of the landslide information values of all the sub-categories present within its respective parameters for each of the polygons.

Though the model presented an accuracy of 95%, however, here mostly soil parameters were taken in detail and very few non-soil parameters were taken into account. Also, the above model showed a high accuracy for a smaller study area; however, with the increase in the study area, the number of polygons will substantially increase and hence might lead to the decrease in the accuracy.

Sharma et al. [7] adopted various fuzzy algebraic functions for the study of vulnerability of landslide in the area of Rumtek-Samdung, East Sikkim.

The study area was further categorized into five different categories by using the LSI; i.e. landslide susceptibility index was calculated from each of the fuzzy operators. Thus, landslide density was then calculated that was the ratio of the number of polygons with landslides to the total geographical area under the categorized susceptibility zone. The resultant accuracy is 96.25%.

Here, for the sub-parameters for the polygons of the causative factor that is taken into consideration, their fuzzy membership value for at least one variable within majority of the polygons will tend to become zero and this may result in giving rather unrealistic information in terms of landslide. Hence, it is observed that for a large number of causative factors and its sub-parameters, effect of the causative factors may not be uniform throughout the study area; hence, the fuzzy operators may not yield the results as expected.

Pradhan et al. [8] proposed a landslide susceptibility model where they have introduced artificial neural network as an approach along with geographic information system and remote sensing data. Ten input parameters pertaining to the landslide of the study area were taken into account to generate a model and assess the vulnerability of the study area to landslides. Based on the causative factors and their assigned weights, the susceptibility of the study area to landslide was calculated. Using the back-propagation training method, the weight of each factor was determined. The accuracy reported was 92.59%.

Neural networks have the ability to allow the target classes to be determined in accordance with their distribution in the corresponding area of the data source, and this also eases the process of integration of remote sensing data.

In this neural network model, the various steps involved in the internal process are complicated and this involves heavy computing load resulting in a long execution time. Thus, the involvement of huge amount of data from the various layers involved in the study area cannot be efficiently processed using artificial neural network programme. Assessment of landslide occurrence potential can be done using the attribute data; however, its events cannot be predicted.

Ahmed [9] made a comparative study of the various models that are used for assessing the landslide susceptibility using OWA: ordered weighted average, AHP: artificial hierarchy process and WLC: weighted linear combination, which are three different geographic information systems, based on multi-criteria decision analysis method. The accuracy of the methods was projected as AHP 89.90%, WLC 83.90% and OWA 87.10%, respectively.

Here, nine different causative factors were taken into consideration as the input parameters. Though the techniques used showed a fair accuracy rate and it is also very useful when dealing with large data/map with ease, however its limitation lies in the fact that these methods also have the influence of the expert's knowledge which hinders the proper selection of the factors affecting the landslide.

Chalkias et al. [10] proposed a model for mapping of landslide susceptibility of the given study area using an integrated approach by using a combination of expert-based and statistical approach. Here, seven causative factors are considered as the input parameters.

Here, statistical analysis along with expert knowledge was used for the study which resulted in an accuracy of 76%. However, total amount of landslide that has already occurred within the area of study was not considered in the data set. Equal amount of significance was given to both expert-based and the statistical models. However, the proposed method did not include the mutual relationship between the causative factors and that the model is only applicable to regional-scale data sets.

Kainthura et al. [11] proposed a model using *K*-means clustering algorithm. They chose the causative factors based on its relevance to the landslide of the area. The parameters which they have identified as the most significant were geomorphology, slope, elevation, rainfall and vegetation.

For defining the levels of rainfall, *K*-means clustering algorithm was used and then the decision tree techniques have been applied to classify the areas that were susceptible to landslide. However, here the accuracy of the technique used has not been projected and only the applicability of the techniques has been shown.

Quan et al. [12] have used artificial neural network (ANN) and analytical hierarchy process (AHP) techniques for the study of vulnerability of the study area to the landslide. Seven causative factors were taken into consideration. A comparative analysis of the given techniques was done to find the most suitable approach that would be most appropriate for selection of only the most relevant landslide parameters that may be applied for valuation of the proposed model. Accuracy 75% was achieved using artificial neural network (ANN) and 58.3% using analytical hierarchy process (AHP). However, soil type and also its geology were reflected as the limitation of these projects as that they did not prove to be very efficient due to the various characteristics associated with it.

Kayastha et al. [13] projected a model for landslide susceptibility map based on analytical hierarchy process (AHP) using eleven causative factors. Here, both the physical and statistical methods were used for the validation of the model developed for mapping of landslide susceptibility for the given study area. The results revealed that there was a strong relationship between the present and the past landslide occurrences that was obtained from the predicted susceptibility levels. However, analytical hierarchy process (AHP) approach is based on the experience, knowledge and judgment of the expert knowledge and these factors highly influence the decision-making.

Moon et al. [14] made a comparative analysis of the methods at a regional scale using bivariate and multivariate statistics. For bivariate statistical approach, weight of evidence (WOE) method was used and logistic regression (LR) was used for multivariate statistical approach.

These statistical analysis approaches along with GIS tool were used to present a model for landslide susceptibility mapping. Here, weight of evidence (WOE) approach was projected as one of the efficient methods for identifying variables that contributed towards occurrence of landslides in the areas under study. Recognition of these important parameters also aided in controlling the landslide processes only for those areas for which sufficient amount of landslide data are available. However, this method failed where complete landslide data was not available whereas the logistic regression model posed a drawback when too many parameters were included.

Zhang et al. [15] presented random forest technique as a new integrated classification method for accessing the areas susceptible to landslide. A total of 34 causative factors were considered for the study which projected an accuracy of 86.10%. This technique not only showed a relatively good response in handling multiple input parameters but also was able to efficiently assess the significance of each of these input parameters that contributed towards the landslide occurrence of the study area.

Though this technique was projected as a model with significantly high accuracy even when it involved a huge number of input parameters however, its limitation lies in the fact that random forest technique is highly influenced by scale, type, precision and number of input factors, and also the number of classifiers [15]. Hence, random forest technique is preferable where there is a large number of input parameters wherein the selection of the most significant parameters can be done with ease (Table 1).

Table 1 A comparative analysis of the techniques used for development of model for landslide susceptibility mapping

Techniques	Study area	Landslide causative factors	Performance measurement method	Prediction accuracy	Scope for improvement
Shannon entropy	Rumtek-Samdung area comprising of around 26 villages	Soil depth, drainage characteristics, depth texture hydraulic conductivity, stoniness, erosional characteristics, lithology, foliation, slope, distance to drainage land cover, distance to road	Information value of different parameters in accordance with the spatial variation and prevalence based on Shannon entropy	95%	Based on small study area and only soil parameters considered. Need to consider a larger study area [6]
Landslide susceptibility index based on various fuzzy operators	Lingdum Village near Ranka Village, East Sikkim	Soil depth, drainage characteristics, depth texture, aspect, other geological parameters, hydraulic conductivity, stoniness, erosion, lithology, foliation, slope, distance to drainage land cover, distance to road	Fuzzy logic-based performance index	92.65%	Fuzzy logic may not yield a better accuracy with the increase in the size of the causative factors and its sub-parameters resulting in unrealistic information in terms of landslide [7]
ANN using back-propagation	Eastern part of Selangor State, Malaysia	Drainage, soil map, land cover, vegetation index (ndvi)	Receiver operating characteristic curve (ROC)	92.59%	Incapability to handle a large data set. The causative factors were taken based on previously existing database [8]

(continued)

Table 1 (continued)

Techniques	Study area	Landslide causative factors	Performance measurement method	Prediction accuracy	Scope for improvement
GIS-based multi-criteria decision using AHP, WLC and OWA	Chittagong metropolitan area, Bangladesh	Distance to drain (m), elevation (m), land cover, vegetation index (ndvi), precipitation, distance to road, slope	Relative operating characteristic curve (ROC)	AHP: 89.90% WLC: 83.90% OWA: 87.10%	The assignment of weights to each factor should be done very carefully as it is based on the expert knowledge. Selection of the causative factors by the experts is based on same parameters used in previous studies of the same area [9]
Combination of expert-based and bivariate statistical methods	Peloponnese, Greece	Peak ground acceleration slope, lithology, mean annual precipitation, elevation, land cover, aspect and slope angle	Receiver operating characteristic curve (ROC)	76%	Integration of expert-based and statistical method showed a subsequent accuracy. Also, the accuracy of the model highly depended on the expert knowledge. Also, the study applies to a medium-scale study area, and for a larger area, the approach may not be able to yield a better result [10]

(continued)

Table 1 (continued)

Techniques	Study area	Landslide causative factors	Performance measurement method	Prediction accuracy	Scope for improvement
Reinforcement learning methods	Uttarkashi district, Uttarakhand	Vegetation, rainfall, geomorphology, slope and elevation	–	–	Memory may pose to be a serious disadvantage. The approach may be comparatively complex as compared to other approaches. Thus, it may be advisable to use the reinforcement learning technique in combination with other techniques which may result in increasing the efficiency of the model [11]
ANN: artificial neural, network, AHP: analytic hierarchy process	Jeju Island, volcanic island, southern coast of Korea	Slope, aspect, geology, soil, rainfall intensity, forest, land cover, weight	Receiver operating characteristic curve (ROC)	Analytic hierarchy process (AHP): 58.3%	Assignment of weights has been done based on subject knowledge, and the assignment of weight plays a vital role yielding the accuracy for mentioned approaches [12]

(continued)

Table 1 (continued)

Techniques	Study area	Landslide causative factors	Performance measurement method	Prediction accuracy	Scope for improvement
Analytical hierarchy process (AHP)	Nepal, area of Tinau watershed	Slope angle, distance from syncline folds, annual rainfall, slope aspect, slope curvature, distance from faults, distance from anticline folds, relative relief, geology, land use, distance from streams	Chi-square test and receiver operating characteristic curve (ROC)	77.54%	Though analytical hierarchy process approach yielded a better accuracy, however, the accuracy of the model highly depends on the expert-based knowledge [13]
Logistic regression and statistical index	Sado Island, Japan	Slope angle, elevation, density of geology, slope aspect, distance to faults, total curvature, drainage density, distance from drainage networks, vegetation index, boundaries, profile curvature, plan curvature, distance to geological boundaries and lithology	Receiver operating characteristic curve (ROC)	81.7%	The categorization of data may serve as one of the disadvantages [14]
Random forest	Zigui–Badong area, China	Elevation, raining data, earthquake magnitude, engineering, rock group, aspect, Vega_PCA1, Mosi_PCA2, Mosi_PCA3, Mosi_PCA1, Vega_PCA2, strata lithology, etc.	Receiver operating characteristic curve (ROC)	86.10%	Type, scale, number of classifiers and precision of input factors affect the classification accuracy of the model. This approach may yield a much better efficiency for a large data set [15]

3 Conclusion

For each of the techniques mentioned above, receiver operating characteristic curves (ROC) have been used for prediction and validation of accuracy. Techniques used for categorization of landslide susceptibility map can be primarily classified into four main categories, namely heuristic, statistical, deterministic and probabilistic techniques.

However, probabilistic and statistical methods are the most preferred ones [3]. These techniques when used in combination with the qualitative techniques have also yielded better results as the expert-based inputs make a significant contribution towards geological terrain features.

For the techniques used above, the selection of the causative factors in most of the cases has been done based on the past study. The accuracy of any landslide susceptibility depends upon the proper selection of the causative factors. However, at the same time, the number of the input parameters taken into account should also be taken into account; for this purpose, proper optimization of the causative factors can be done so that those factors having the higher significance towards the occurrence of the landslides can be done.

Since landslide susceptibility mapping of an area and its risk assessment involve a huge volume of data, how efficiently can the model handle these huge volumes of data has not been addressed.

Use of multiple factors (interdependency of causative factors) is done to yield a better accuracy. Also, triggering factors highly influence the landslide conditioning factors. Hence, there arises the need to check the level of influence of the triggering factors on the landslide conditioning factors (interdependency of causative factors). This issue has been checked by only taking few parameters into consideration. However, a larger number of parameters have not been addressed.

Hence, machine learning is gaining much more popularity as these techniques have the capability to independently adapt, when exposed to new set of data. They also possess the quality to learn from the previous computations to yield much efficient and reliable result. Also, they have the capability to handle large and complex data and also efficient handling of incomplete data [15].

References

1. Sujatha ER, Rajamanickam GV, Pichaimani K (2012) Landslide susceptibility analysis using probabilistic certainty factor approach: a case study on Tevankarai stream watershed, India. J Earth Syst Sci. 121. https://doi.org/10.1007/s12040-012-0230-6
2. Ali Y (2008) GIS-based landslide susceptibility mapping using analytical hierarchy process and bivariate statistics in Ardesen (Turkey): comparisons of results and confirmations. Catena 72:1–12. https://doi.org/10.1016/j.catena.2007.01.003
3. Brabb EE (1985) Innovative approaches to landslide hazard and risk mapping. In: International landslide symposium proceedings. https://pubs.er.usgs.gov

4. Vahidnia M, Alesheikh A, Alimohammadi A, Hosseinali F (2009) Landslide hazard zonation using quantitative methods in GIS. IJCE 7(3):176–189. http://ijce.iust.ac.ir/article-1-289-en.html
5. Korup O, Stolle A (2014) Landslide prediction from machine learning. Geology today, vol 30, no 1. Wiley, The Geologists' Association & The Geological Society of London
6. Sharma L, Patel N, Ghose M, Debnath P (2015) Development and application of Shannon's entropy integrated information value model for landslide susceptibility assessment and zonation in Sikkim Himalayas in India. Nat Hazards 75. https://doi.org/10.1007/s11069-014-1378-y
7. Sharma LP, Patel N, Ghose MK et al (2013) Synergistic application of fuzzy logic and geo-informatics for landslide vulnerability zonation—a case study in Sikkim Himalayas, India. Appl Geomat 5(4):271–284. https://doi.org/10.1007/s12518-013-0115-
8. Pradhan B (2011) An assessment of the use of an advanced neural network model with five different training strategies for the preparation of landslide susceptibility maps. J Data Sci 9:65–81
9. Ahmed B (2014) Landslide susceptibility mapping using multi-criteria evaluation techniques in Chittagong Metropolitan Area, Bangladesh
10. Chalkias C, Polykretis C, Ferentinou M, Karymbalis E (2016) Integrating expert knowledge with statistical analysis for landslide susceptibility assessment at regional scale. Geosciences 6:14
11. Kainthura P, Singh V, Gupta S (2015) Gis based model for monitoring and predition of landslide susceptibility. In: 1st International conference on next generation computing technologies (NGCT), pp 584–587. IEEE
12. Quan HC, Lee BG (2012) GIS-based landslide susceptibility mapping using analytic hierarchy process and artificial neural network in Jeju (Korea). KSCE J Civ Eng 16(7):1258–1266. https://doi.org/10.1007/s12205-012-1242-0
13. Kayastha P et al (2013) Application of the analytical hierarchy process (AHP) for landslide susceptibility mapping: a case study from the Tinau watershed, west Nepal. Comput Geosci 52:398–408
14. Dou J et al. (2015) Optimization of causative factors for landslide susceptibility evaluation using remote sensing and GIS data in parts of Niigata, Japan. http://dx.doi.org/10.1371/journal.pone.0133262
15. Zhang K et al. (2017) The assessment of landslide susceptibility mapping using random forest and decision tree methods in the three gorges reservoir area, China. Springer, Berlin Heidelberg

Technology Enabled Smart Efficiency Parking System (TESEPS)

Amiyangshu Nath, Himshekhar Nr. Konwar, Kapil Kumar and Md Ruhul Islam

Abstract Transportation is the key infrastructure of a country. A country's economy status depends upon how well the country is served by its roads, railways, airports, ports, pipelines and shipping. The rate at which a country's economy grows is very closely linked to the rate at which the transport sector grows. As road transport gives personal mobility to persons, the vehicle ownership rate has been increasing at a fast rate round the world. Due to increase in car ownership, the problem of parking is becoming more and more acute day by day (Singh and Sharma in application of advanced parking management system techniques—a case study. Int Organ Sci Res J Mech Civil Eng 3(2):24–28, 2012, [1]). Parking space utilization is one of the major problems in crowded cities. Even after coming up with regulations and dedicated parking spaces, with the number of increasing vehicles on road each day, it is hardly making any difference in the existing vehicle parking problems. Yes, we see a huge improvement in the efficiency of space utilization while parking the vehicles, but there is more that can be done to improve it even further without building an entirely new parking space. Even in the most efficient vehicle parking spaces, where the vehicles are parked in a horizontal row with only a little space (for the driver to get inside/outside of the vehicle) between each vehicle, the little space between each vehicle collectively makes up a lot of space that is wasted in the name of driver's comfort to get inside/outside the vehicle. This project aims toward developing a solution to eliminate these small spaces between each vehicle without compromising with the driver's comfort in getting inside/outside the vehicle, in fact offering more convenience in doing so and truly making an efficient use of the parking space. There are two types of car parking systems: automated and traditional. In the long term,

A. Nath · H. Nr. Konwar · K. Kumar · Md. R. Islam (✉)
Department of Computer Science and Engineering, Sikkim Manipal Institute of Technology, Sikkim Manipal University, Majhitar, Sikkim, India
e-mail: ruhulislam786@gmail.com

A. Nath
e-mail: amiyangshu1@gmail.com

H. Nr. Konwar
e-mail: himshekhark@gmail.com

K. Kumar
e-mail: ktomer1998@gmail.com

H. K. D. Sarma et al. (eds.), *Trends in Communication, Cloud, and Big Data*, Lecture Notes in Networks and Systems 99, https://doi.org/10.1007/978-981-15-1624-5_14

automated car parking systems are more cost effective when compared to traditional parking garages. IoT is the best option for smart parking system (Osmani et al. in research paper on smart city parking system. Int J Adv Res Innov Ideas Educ 2(3), 2016, [2]). This project incorporates IoT level 1 technology and uses the concepts of bi-directional rotating conveyor belts using DC motors and Arduino microcontroller, NFC tagging and automation. This report depicts the model of the entire system and can be used to build the physical prototype and also build the technology over it.

Keywords Parking space utilization · IoT · IoT level 1 · Arduino · DC motor · NFC · Automation

1 Introduction

Every driver in this country, sometime in their lives, has faced vehicle parking problems. Finding a parking space in most metropolitan areas, especially during the rush hours, is difficult for drivers. The difficulty arises from not knowing where the available spaces may be at that time; even if known, many vehicles may pursue very limited parking spaces to cause serious traffic congestion [3].

In crowded cities, this is an everyday situation that the drivers need to encounter. On one side, drivers complain that there is not much parking space in the public areas and that they have to park their vehicles in a risky position (i.e., very close to adjacent vehicles in a row) that increases chances of minor scratches and clashing of the vehicles with each other, and on the other side, the Government bodies and responsible authorities pressurize the local authorities to waste as little space as possible in the parking areas and utilize every inch of the parking space. The views of both the parties here are colliding, they will continue to collide until a middle ground is found where the drivers feel safe that their vehicle is out of risk of collisions, and the Government is successful in using every inch of the parking spaces. Technology has solved even the unthinkable problems and is now even ready to send people to Mars but cannot anything be done for these existing parking problems? This project outlines one of those million solutions that can prove to find this middle ground among both the parties using IoT level 1 technology.

2 Problem Definition

Crowded cities have enough problems already, and the increasing rate of vehicles on road each day and lesser parking spaces add on to it. Even in the most efficient and safe parking spaces, it does not have a 100% space utilization. In the most efficient parking spaces, where the vehicles are parked side-by-side in a row, it can be still found that there is some space between adjacent vehicles (for driver's convenience in getting in/out of the vehicle) which collectively makes up a lot of potential parking space wasted just like that. On top of that, there is this never-ending war between vehicle drivers and the authorities about this problem—drivers want more individual space for the vehicle for its safety (from scratches/collisions), and authorities want lesser amount of space wasted.

3 Proposed Solution Strategy

This project uses level 1 IoT technology composed of bi-directional DC motor movements, conveyor belt technology, centralized Arduino-based controlling and NFC tagging. An extended conveyor belt will be placed in individual parking spaces which will be used to carry the vehicle on it and pull it into the actual parking space. So, this way the vehicles can be placed at a very close proximity from each other in the parking space. Also, when driver would like to take the vehicle away, the vehicle will be pulled out from the actual parking space on the conveyor belt itself to the extended area, and the driver can easily drive away the vehicle without worrying about any collisions. Also, to keep a track of the vehicle and vehicle owner's identity, NFC tagging is used. The NFC tagging will not only act as a trigger to the movement of the conveyor belt but also will be able to uniquely identify every driver and his/her vehicle.

The components incorporated in the project are

(i) Arduino
(ii) Motor driver
(iii) DC motor
(iv) Bread board
(v) Jumper wires
(vi) NFC tag cards
(vii) Conveyor belt model.

Figure 1 shows the physical design layout of the system. The conveyor belt pulls in and out the vehicles on it to and from the parking slot, respectively.

4 Design Strategy for the Solution

The above flowchart clearly depicts the various actions taking place in the system and the conditions responsible for the actions. It also serves the purpose of giving a clear picture of the working of the system. The detection of NFC card here acts as a trigger to the rotation of the conveyor belt in both inward and outward directions (Figs. 2, 3 and 4).

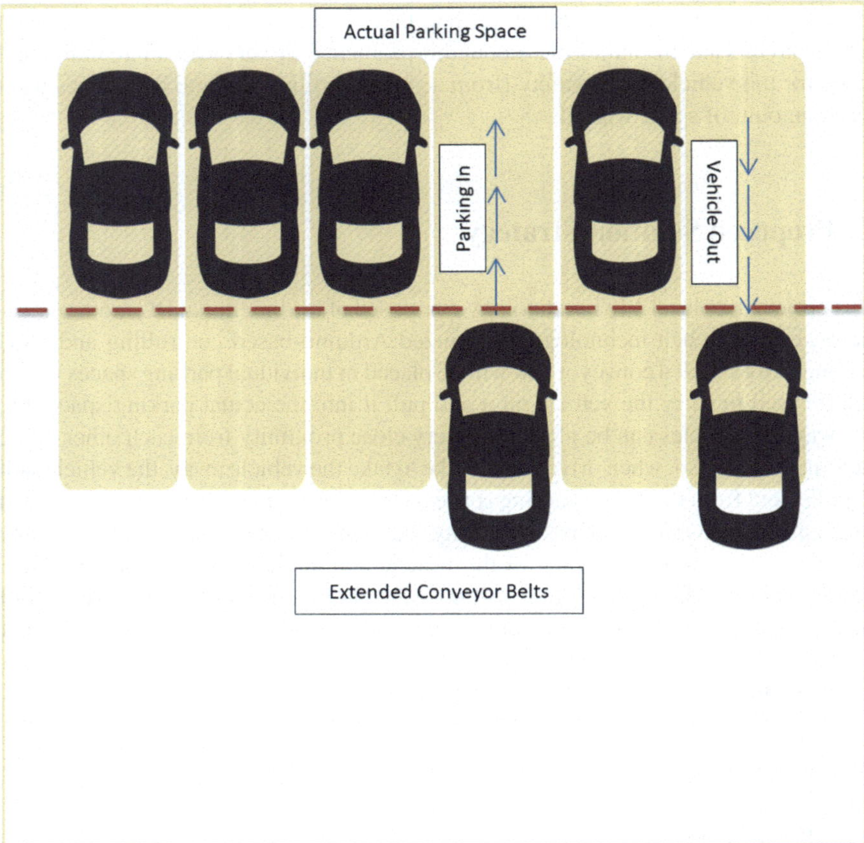

Fig. 1 Design layout

5 Working Principle

5.1 Arduino UNO

An Arduino is originally a microcontroller-based kit. It can be purchased and used or can be assembled and made at home using the components, owing to its open-source hardware features. Controlling and operating multiple devices and communicating between those are possible using this kit. These days many people try to use the Adriano because it makes things easier due to the simplified version of C++ and the already made Arduino microcontroller that you can programme, erase and reprogramme at any given time [4].

The Arduino is the central coordination system in this project responsible for all the movements and activities going on in the project (Fig. 5).

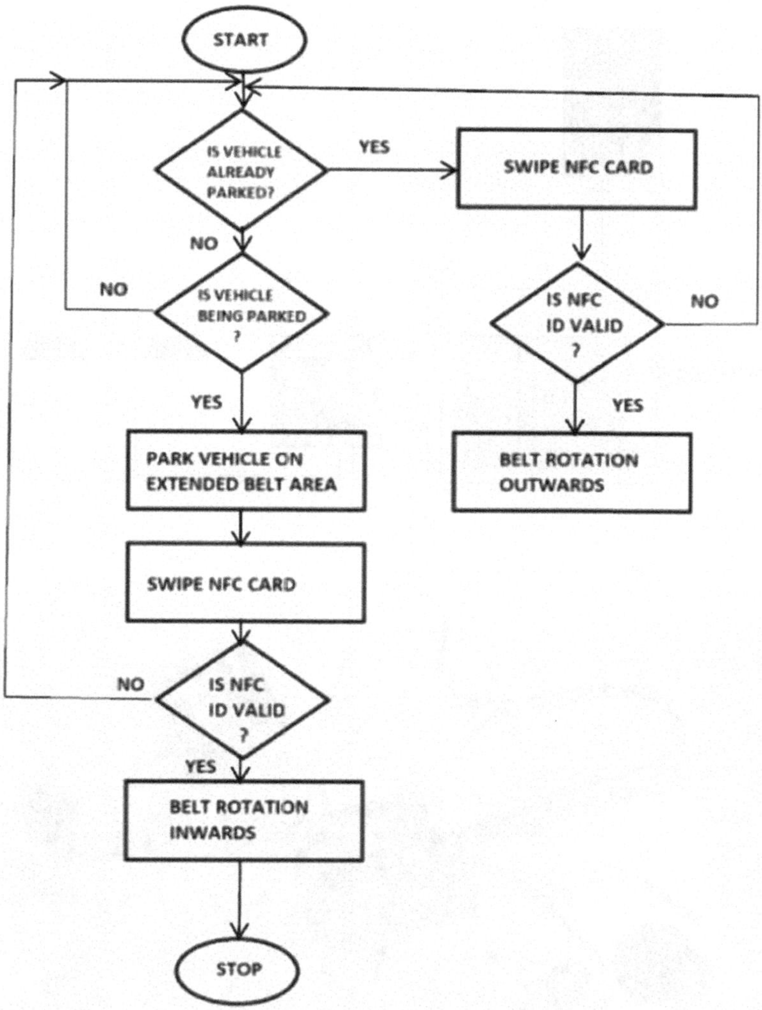

Fig. 2 Flowchart cf solution strategy

5.2 *DC Motors*

A DC motor in simple words is a device that converts electrical energy (direct current system) into mechanical energy [5].

In this project, DC motors are used to run the conveyor belt in forward and backward directions for carrying the vehicle to and out of the parking slot [6] (Fig. 6).

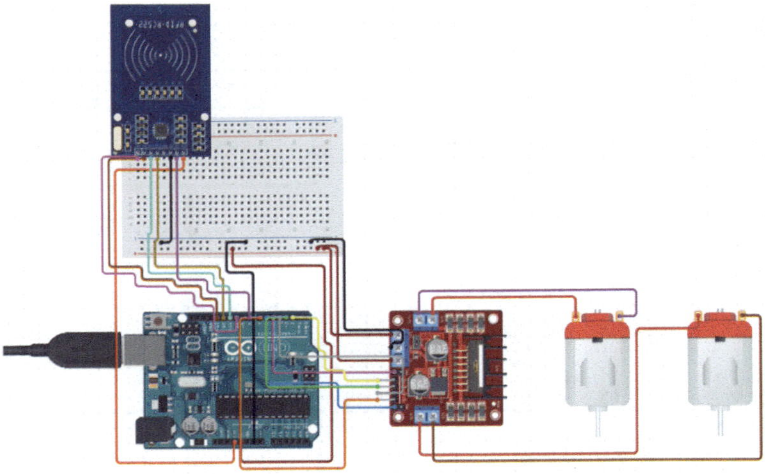

Fig. 3 Circuit diagram of solution strategy

Fig. 4 Real photograph of project circuit

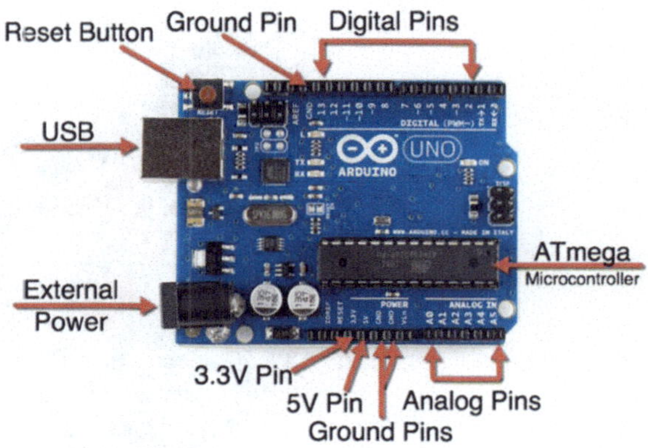

Fig. 5 An Arduino UNO [4]

Fig. 6 DC motor [5]

5.3 *L298 Motor Driver H-Bridge Module*

L298 Motor Driver Module is an integrated monolithic circuit in a 15-lead Multiwatt and PowerSO20 packages. H-Bridges are typically used in controlling motors speed and direction. An H-Bridge is a circuit that can drive a current in either polarity and be controlled by pulse width modulation (PWM). Pulse width modulation is a means in controlling the duration of an electronic pulse [7, 8]. In this project, L298 Motor Driver is used to control the DC motor operations and its movement in both the directions (Fig. 7).

5.4 *NFC V3 Module*

NFC is a set of communication protocols that enables two electronic devices to establish communication by bringing them within 4 cm of each other. In this project, NFC cards are being used for authorization purpose of the vehicle driver to ensure safety of the vehicle [9] (Fig. 8).

Fig. 7 L298 Motor Driver module [6]

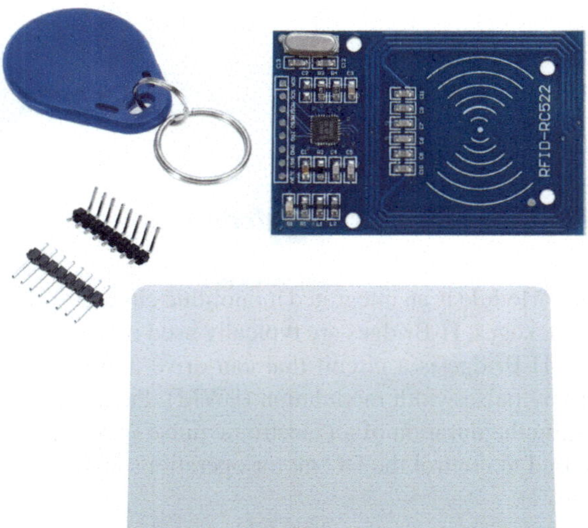

Fig. 8 NFC module (card and chip)

6 Implementation and Result

See Table 1 and Fig. 9.

7 Conclusion

This project has successfully cut down the wastage of the free parking space in the parking lots, has established an efficient way to utilize almost every inch of the parking space for its purpose and in the way has also established convenience for the drivers to park their cars without being tensed about scratches and minor clashes. It has been able to finally end the long-drawn dispute between the drivers, wanting more space for parking for convenience and safety, and the authority, wanting to waste as less space as possible. This system perfectly reaches the middle ground on

Table 1 Implementation and results

Sl. no.	Test case	Response
1	Testing Arduino by blinking in-built LED	LED blinking Adriano working message
2	Testing NFC V3 module	NFC read message (NFC ID detection)
3	Testing DC motors	Mechanical movement (circular rotation in both clockwise and anti-clockwise direction.)
4	Testing motor driver using DC motors	Mechanical movement (circular rotation) of DC motors

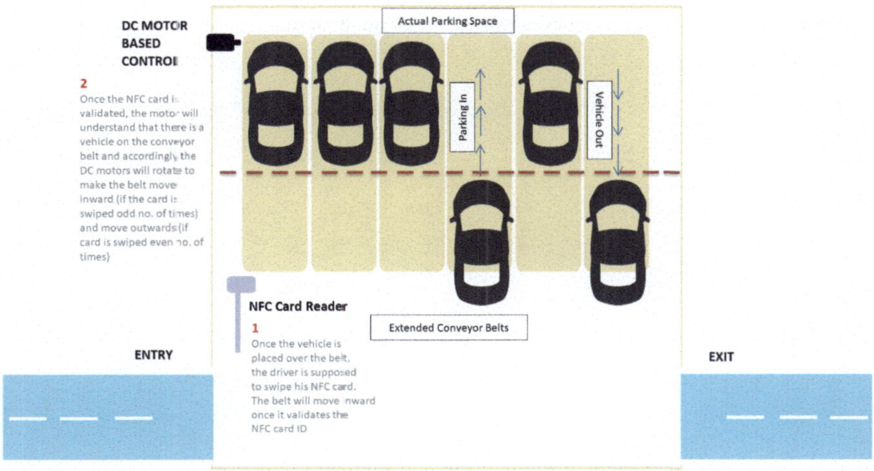

Fig. 9 Final model of the proposed system

both the parties' wants by offering more convenience in parking and ensuring safety of the vehicle from accidents to the drivers and at the same time use every inch of the parking space for its purpose.

This model can also work as the base model for newer technologies to be built over it. The project has a huge scope for lot of applications in the parking industry as well as other industries requiring mechanical movement and transport of material in and out of a compound. More advanced technology like IoT, cloud computing and AI can be implemented in this system to make a lot of processes in the system automated and enable remote controlling of the system too.

References

1. Singh ES, Sharma U (2012) Application of advanced parking management system techniques_a case study. IOSRJMCE 3(2), 24–28. ISSN: 2278-1684. www.iosrjournals.org
2. Osmani AH, Gawade A, Nikam M, Wavare S (2016) Research paper on smart city parking system (Under the guidance of prof. Kalpana Kadam). IJARIIE 2(3). ISSN(O)-2395-4396
3. Wang H, He W A reservation-based smart parking system. http://cse.unl.edu/~byrav/INFOCOM2011/workshops/papers/p701-wang.pdf
4. Badamasi YA (2014) The working principle of an Arduino. In: 11th international conference on electronics, computer and computation (ICECCO). Nigerian Turkish Nile University, Abuja, Nigeria
5. Working or operating principle of DC motor. https://www.electrical4u.com/working-or-operating-principle-of-dc-motor/
6. Reichenstein7 Arduino modules—L298N dual H-bridge motor controller. https://www.instructables.com/id/Arduino-Modules-L298N-Dual-H-Bridge-Motor-Controll/
7. Taher GA, Howlader Y, Rabbi MA, Touqir FA (2014) Automation of material handling with bucket elevator and belt conveyor. Int J Sci Res Publ 4(3). ISSN 2250-3153
8. How to control a DC motor with an Arduino. https://www.allaboutcircuits.com/projects/control-a-motor-with-an-arduino/
9. https://www.tutorialspoint.com/arduino/arduino_dc_motor.htm

CLEAN Algorithm for Imaging Quality Enhancement

S. N. Sur, S. Bera, A. Anand and S. Rai

Abstract In ITS, radar plays an important role. Detection of the target and characterization is very important. Similarly, in dynamic channel condition, target image quality degrades considerable due to sidelobe enhancement and defocusing. The clean algorithm is introduced to eliminate sidelobes and significantly improves the target detection performance of radar signals. The experimental results demonstrate the effectiveness of the CLEAN algorithm.

Keywords Clean algorithm · RADAR · Inverse synthetic aperture radar (ISAR) · Range · Correlation · Spectrum

1 Introduction

Automobile radar is one of highly growing industries due to increase in the heavy traffic worldwide. Along with the increase with the on-road traffic, safety has become very important. Also, traffic congestion is another aspect to look into. Hence, the introduction of intelligent transportation system (ITS) in highway network is of utmost significance [1]. One of the solutions to the above problems is remote sensing [2, 3] the different obstacles and act accordingly. Remote sensing is again a challenging job under ITS due to highly dynamic channel condition. In dynamic channel situation, the nearby or adjacent targets suffer from increase in sidelobe and thus overlapping on each other causing target peak misallocation [4, 5]. So, to meet the demand, very high-resolution imaging radar needs to be developed and this paper deals with the development of such high-resolution imaging radar for vehicular application so as to recognize and classify targets or obstacles clearly and improving the on-road traffic conditions [6–8].

The requirement of high-resolution radar image so as to identify the targets clearly a technique known as CLEAN [4, 9, 10] may be employed which has the capability

S. N. Sur (✉) · S. Bera · A. Anand · S. Rai
Sikkim Manipal Institute of Technology, Sikkim Manipal University, Majitar, Rangpo, Sikkim 737136, India
e-mail: samar.sur@gmail.com

H. K. D. Sarma et al. (eds.), *Trends in Communication, Cloud, and Big Data*, Lecture Notes in Networks and Systems 99, https://doi.org/10.1007/978-981-15-1624-5_15

of reducing the sidelobes enhancement thus lowering the chances to target overlapping. Hence, the application of this algorithm will lead to proper localization and classification of the closely spaced vehicular targets so as to meet the challenge of heavy traffic. In this paper, authors have introduced a hybrid processing scheme for a polyphase coded spread spectrum system by combining the matched filter and coherent clean algorithm in order to enhance the performance of the radar.

2 Clean Algorithm

In this paper, the CLEAN algorithms are used to improve the image quality of polyphase coded spread spectrum ISAR radar. The practical result demonstrates the effectiveness of the CLEAN algorithm. Here, signal processing methods, like cross correlation, FFTs, and CLEAN algorithm on target images, are incorporated to improve the radar system performance.

The algorithm, as in Fig. 1, initially brightest spot in the image map (D) has been identified. The location and intensity information for the brightest spot are retained in a clean map (G). Once the brightest spot is removed from the image map (D), the procedure is repeated with the next brightest source in the image map until some predetermined condition has been reached. For the spot identification, peak detection or blind spot detection algorithm is used to prepare a hot spot imaging.

3 Experimental Results

Here, for the sensing purpose, polyphase coded signal has been used and has 100 MHz bandwidth. Arbitrary wave generator (AWG) is used to generate IF signal over 70 MHz. For increasing the radar range resolution RF frequency swept over the frequency rage 1–2 GHz. For the target characterization toy car has been placed over a 2-axis positioner. The target has the dimension of 0.6 m (cross range) $\times$ 0.9 (range). For range cross range 2D imaging we have rotated the target in the azimuth direction. The hart of the receiver is vector signal analyzer (VSA), which is a programmable device. All the receiver signal processing algorithms have been implemented in the VSA. The down-converted signal is used to extract the target information using VSA.

The radar signal processing (RSP) algorithm includes a correlation process for ranging information followed by 2D FFT for cross range information. The CLEAN algorithm has been applied to improve the image quality.

Figure 2 represents image of the target without background cancellation. As in the figure, the target is detected but with back scattered false target. This additional hot spot results in false detection and needs to be removed.

Whereas, Fig. 3 represents image of the target after background cancellation. As in figure, background cancellation effectively able to remove the back scatters.

Effectiveness of the CLEAN algorithm is depicted in Fig. 4. By comparing

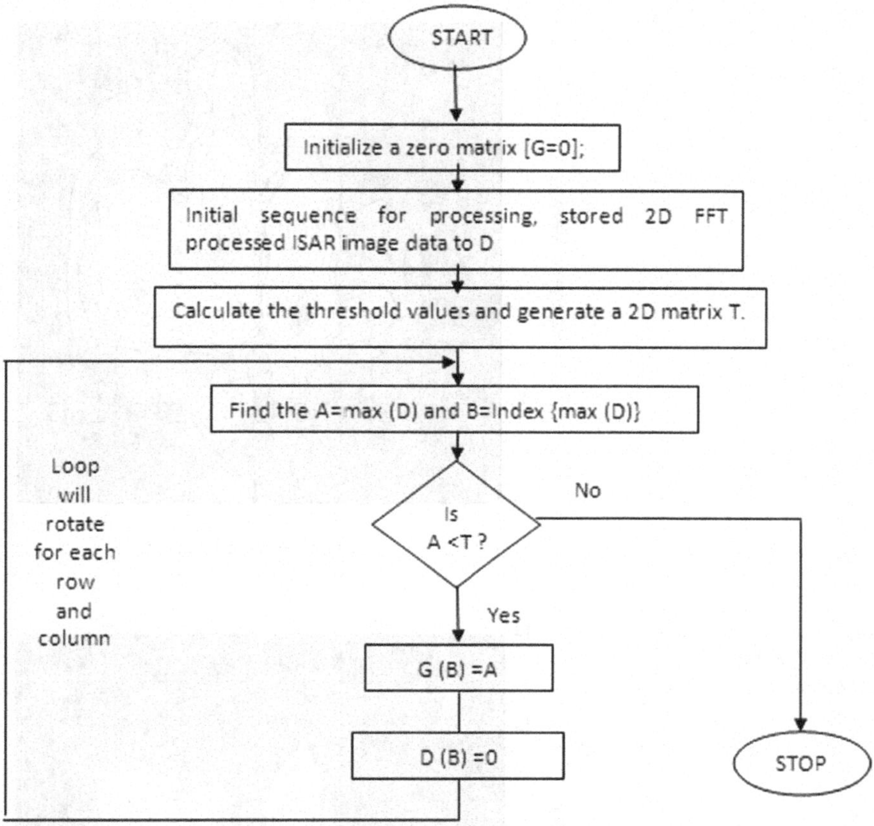

Fig. 1 CLEAN algorithm for image cleaning

Figs. 3 and 4 one can easily figure out the impact of the CLEAN algorithm. CLEAN algorithm helps to create more distinct image and thereby improves the image quality.

4 Conclusion

This paper demonstrates the CLEAN algorithm and its effectiveness for creating hot spot image of the target. In critically danse conditions or during the imaging of the contiguous target, the image quality degrades due to sidelobe enhancement. And in this regard CLEAN algorithm is a very good tool to improve the image quality. As depicted in the result above it can be seen that the efficient use of CLEAN algorithm provides significant enhancement in the performance of the system.

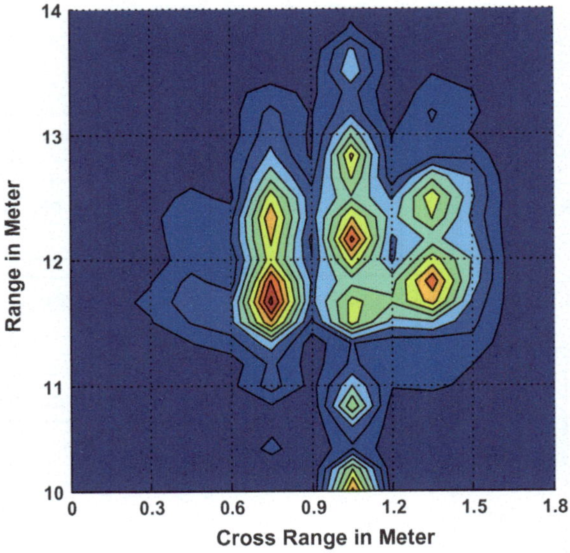

Fig. 2 Target (car) image before background cancellation

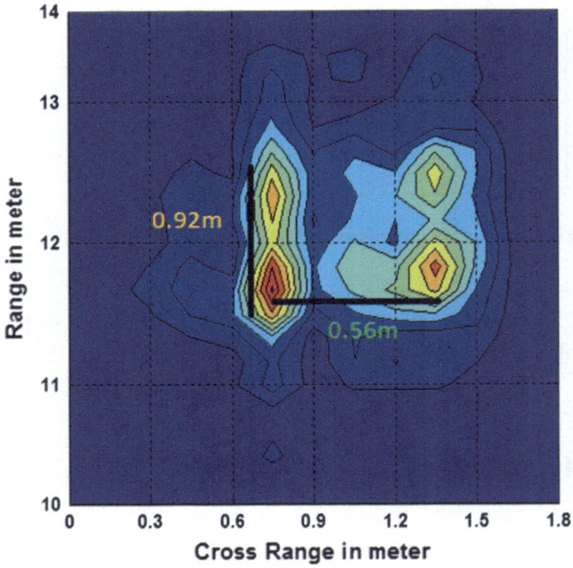

Fig. 3 Target (car) image after background cancellation

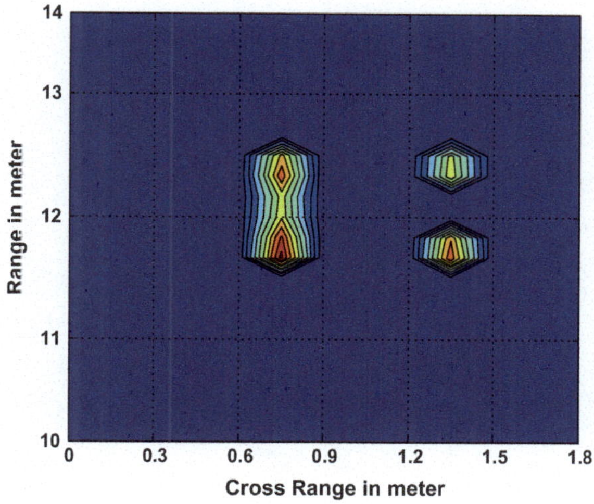

Fig. 4 Cleaned target response

References

1. EU eSafety (2002) Final report of the eSafety working group on road safety, pp 7–39
2. Cooke T, Martorella M, Haywood B, Gibbins D (2006) Use of 3D ship scatterer models from ISAR image sequences for target recognition. Digit Signal Proc 16(5):523–532
3. Kim K-T, Seo D-K, Kim H-T (2005) Efficient classification of ISAR images. IEEE Trans Antennas Propag 53(5):1611–1621
4. Bose Ranjan (2011) Lean CLEAN: deconvolution algorithm for radar imaging of contiguous targets. IEEE Trans Aerosp Electron Syst 47(3):2190–2199
5. Tsao J, Steinberg BD (1988) Reduction of sidelobe and speckle artifacts in microwave imaging: the CLEAN technique. IEEE Trans Antennas Prop 36:543–556
6. Ghaemi H, Börner T, Galletti M, Viberg M, Gekat F (2009) CLEAN technique in strip-map SAR for high-quality imaging. In: IEEE aerospace conference. Big Sky, MT, USA, 07 Mar 2009–14 Mar 2009
7. Foreman TL (2010) Application of the CLEAN detector to low signal to noise ratio targets. In: 2010 IEEE international radar conference (9th), Arlington, Virginia, 10–14 May 2010
8. Almutiry M, Monte LL, Wicks MC (2017) Extraction of weak scatterer features based on multipath exploitation in radar imagery. Int J Antennas Prop 2017:1–13. Article ID 5847872
9. Hogbom JA (1974) Aperture synthesis with non-regular distribution of interferometer baselines. Astron Astrophys Suppl 15:417–426
10. Sur SN, Bera S, Singh AK, Bera R (2012) Improvement in target detection and image. formation for ISAR system using CLEAN algorithm. In: CODIS, pp 429–432. https://doi.org/10.1109/codis.2012.6422230

Context-Aware Song Recommendation System

Sudipta Chakrabarty, Sangeeta Banik, Md Ruhul Islam and Hiren Kumar Deva Sarma

Abstract The paper presents an intelligent method to generate list of songs online for listening, download according to age factor of users. Context-aware is a process used in smartphone or computer system that can sense their physical environment and adopt their behavior correspondingly. It is a Web-based application software that recommended the different songs depends upon the listener choice based on their age group from the music library and also classify the unknown songs in the same cluster depends on the review of user.

Keywords Context-aware · Recommendation system · Web-based application · Music library · Users' feedback

1 Introduction

Context-aware song recommendation system is a Web-based application where a user can provide their date of birth, and the system has to fetch the song according to their age group from the music library and also classify the unknown songs in the same cluster depends on the user feedback. The system generates the suitable songs for the different users. Primarily, the user has to register themselves and create their own profile after that the user can easily login to their account and view the songs. The user can divide into two categories—administrator and user. There have some basic characteristics of administrator and user. Administrator is the super-user of this Web site, can view all user details. Administrator has the access to master forms for

S. Chakrabarty (✉)
Department of MCA, Techno India, Salt Lake, Kolkata, West Bengal, India
e-mail: chakrabarty.sudipta@gmail.com

S. Banik
Department of CSE, SNU, DG Block, New Town, Kolkata, West Bengal, India

M. R. Islam
Department of CSE, Sikkim Manipal Institute of Technology, Majhitar, Rangpo, Sikkim, India

H. K. D. Sarma
Department of IT, SMIT, Majhitar, Rangpo, Sikkim, India

H. K. D. Sarma et al. (eds.), *Trends in Communication, Cloud, and Big Data*, Lecture Notes in Networks and Systems 99, https://doi.org/10.1007/978-981-15-1624-5_16

the purpose of insertion, updating and deletion of databases and viewing the user details and update his login credentials. Administrator can forward songs to the user and upload the songs according to the different age group's choice. User can view the Web site and register him/herself to login into the Web site. User can view and update his/her profile, fetch songs according to age from the database, can play and download songs and review song.

2 Related Works

The most common event for music composers is to generate versatile rhythms. Paper [1] introduces how to generate musical pattern using pattern recognition technique. This technique used to resolve the tonal context which is restricted. Paper [2, 3] introduces pattern recognition of music by mathematical expressions. Paper [4, 5] describes how genetic algorithm used in the field of music composition. In paper [6], the genetic algorithm operator introduces offspring rhythm from a pre-defined set of rhythms. In paper [7, 8], the genetic algorithm is used to generate the realistic drum-set rhythm. Some researcher presents a system in paper [9] where the median filter helps to recognize the pattern in the Indian music and proposed the three measurements of generating rhythm complexity. In paper [10, 11], some researchers developed the object-oriented concept of the musical pattern recognition and rhythmic features retrieval. Some papers describe the implementation of musical bang of tempo [12, 13].

It can be improved the quality of music using genetic algorithm through finding the fittest rhythm structure. In paper [14], a learning tool for music rhythm learner is introduced. This tool is used to develop an adaptable music rhythm from a set of rhythms that contain more melody and tempo. Music is an art of combining sounds. Computational musicology is a developing field that depends on computer science. A lot of research concerned with Indian music recommendation is already present by various researchers [15]. Unified modeling language is introduced to represent the song origin [16]. Music recommendation system is one of the highly intelligent tools that differentiate songs at different time slots in a whole day [17]. Paper [18] proposed a method that classifies the song patterns through the coefficient of variance.

3 Proposed Work

The process of the proposed work is given below:

Part 1: If the song belongs to our music library, then the following steps are allowing:

Step 1: Open the Web site "Context-Aware Song Recommendation System."

Step 2: Provide the required details to register from the register page and click on the "submit" button. The details get stored in "music lover" database.
Step 3: The following options are available in our Web site:

- Login as User
- Login as Admin.

Step 4: Next provide the registered email-id and password to Sign-In into the Web site. The validity of the details is checked using data stored in "music lover" database.
Step 5: In the user page click on "Know your Song."
Step 6: Enter the proper date of birth and click on "Show Me."
Step 7: According to the date of birth list of songs is generated.
Step 8: The song name field from the "music" table is matched with the date from and date to name field of the "music" table to get the best-fitted song according to the age group by the help of date of birth attribute.
Step 9: Next on clicking on the song name, a new page is opened.
Step 10: Click on the "player" to play the song and stop the song. By clicking the download button, it will be downloaded and by clicking the sign out button it will also be stopped.
Step 11: User can also review songs according to their choice.

Part 2: If the song does not belongs to our music library that means for any uncommon songs the following steps are allow:

Step 1: Open the WI_FTP application from any android smartphone.
Step 2: Two options are available on the WI_FTP - SEND and RECEIVE.
Step 3: If we want to send a song, select SEND option, then choose a song from the phone memory.
Step 4: Then connect the RECEIVER to the SENDER's hotspot.
Step 5: Receiver needs to download the song and play the song and give the feedback to the admin using a Google form and depends on the feedback; we classify the song on the particular age group.

The workflow to create the WI_FTP as follows:

Step 1: Create android project (one module library and one app demo).
Step 2: Module consisting of file picker, cache image downloader, hotspot creator and other support libraries.
Step 3: Demo consisting of app resources layout files icon background and java files to provide the functionality to call other classes from library.
Step 4: Finally, manifest file to manage all the screens and declare the permission required by app like read internal storage, access Internet, etc.

To maintain the work properly it requires a huge database of different types of songs using the help of cloud connection by Cloud Server or Oracle Server 10G (without cloud architecture).

Steps of database connection by Oracle 10G

Step 1: Open ORACLE Express Edition showing in Fig. 1.

Fig. 1 Open ORACLE express edition

Step 2: Login with user id and password.
Step 3: Select create tables from Oracle home page.
Step 4: Table will be created after specify following details is shown in Fig. 2.
Step 5: Use below code to connect with cloud.

```
public Connection conClass() {
    try {

        Class.forName("oracle.jdbc.driver.OracleDriver");
        c=DriverManager.getConnection("jdbc:oracle:thin:@localhost:1521:x
e","system","system");

    }catch(Exception e) {System.out.println(e);}
    return c;
}
```

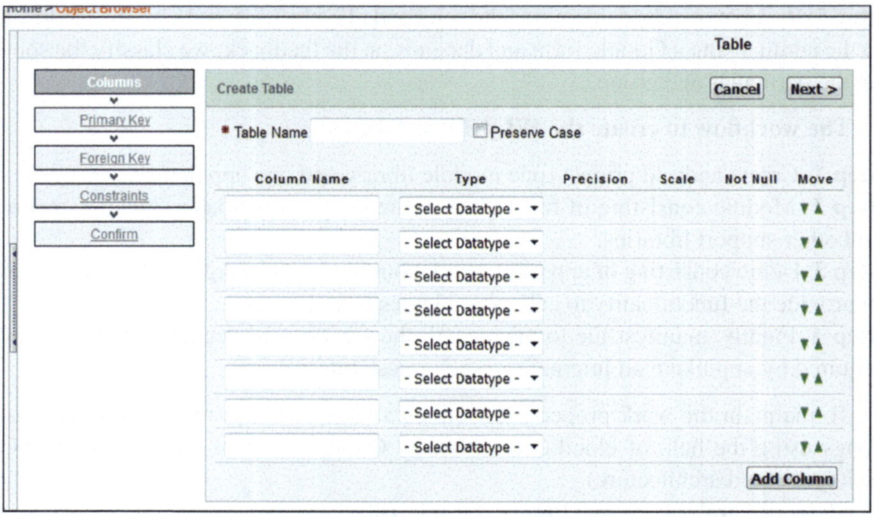

Fig. 2 Table will be created after specify the following details

Steps of Cloud Connection

In this project, we are using the cloud technology so here is some steps to follow to connect with the Cloud Server.

Step 1: Opening phpMyAdmin page is showing in Fig. 3.
Step 2: Login with user name and password.
Step 3: Creating the database and tables of the database is showing in Fig. 4.
Step 4: Connect with the local host with IP address.
Step 5: Use below code to connect with cloud.

```
public Connection conClass() {
    try {
    Class.forName("com.mysql.jdbc.Driver"); c=DriverMan
ager.getConnection("jdbc:mysql://abccollaborations.com/ardent1","user_id","pa
ssword");

    }catch(Exception e) {System.out.println(e);}
    return c;
}
```

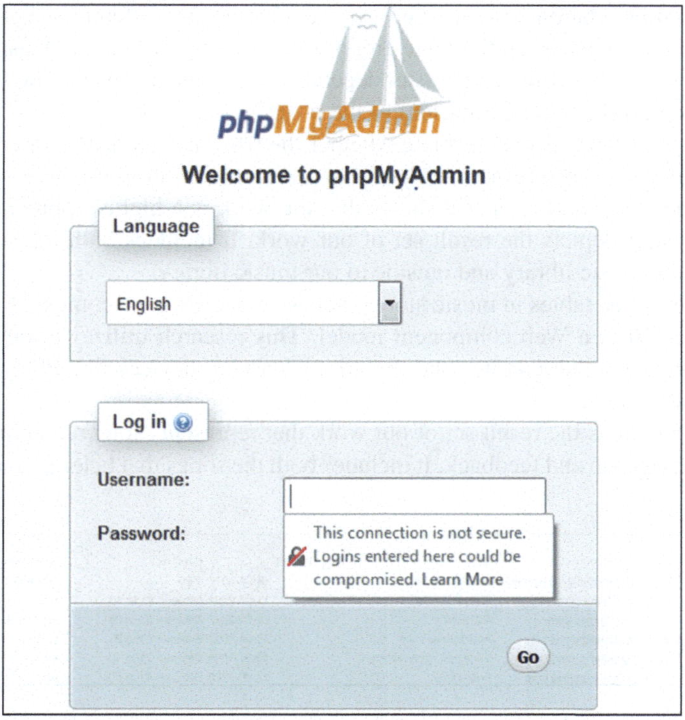

Fig. 3 Opening phpMyAdmin page

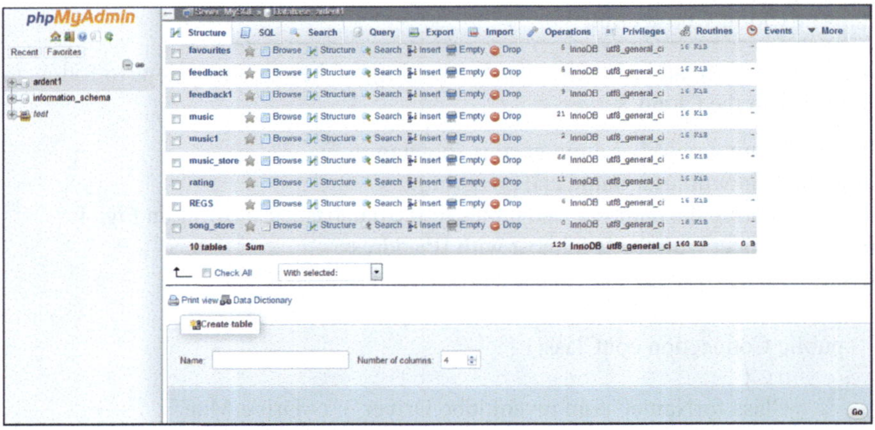

Fig. 4 Creating the database and tables of the database

4 Result Set Analysis

The primary aim of this paper is to a way of listing music in a new manner. User can fetch the song from music library by providing their individual date of birth. This is a song recommendation system where we can classify the different songs depends on the user choice based on their age group from our music library and also classify the unknown songs in the same cluster depends on the user feedback. The following diagram depicts the overall flowchart of the project.

The main purpose is to determine whether the work is technically, operationally and economically feasible or not. Algorithm is selectively choosing the correct song from all the songs and it clearly shows that the work is a highly appreciable. The following chart depicts the result set of our work. It includes both the songs that belong to our music library and outside to our music library.

There are three tables in music store database namely—music, music lover, user login details. It is a Web component model. This research utilizes to develop the learning objective that can be used in various learning devices like PDA, wireless sensors and servers.

Figure 5 depicts the result set of our work that represents song recommendation based on age group and feedback. It includes both the songs that belong to our music

A	B	C	D	E	F
Timestamp	Email address	Name	Date Of Birth	Feedback	Song Name
5-11-2017 0:53:31	arunavadutta10@gmail.co	Arunava Dutta	06-04-2017	1-20	Jingle Bell
5-11-2017 1:05:17	smn.banik@rediff.com	Suman Banik	11-02-2001	between 21 to 40	Naina
5-12-2017 1:03:06	avik@gmail.com	Avik Sen	24-05-2017	Between 61 to 80	AA Chal ke tujhe
5-12-2017 1:03:39	amitpaul@gmail.com	Amit Kumar Paul	26-05-2017	Between 61 to 80	Chingari
5-12-2017 1:11:52	sandipan@gmail.com	Sandipan Chakraborty	13-04-1996	Between 21 to 40	Kaabil
5-12-2017 8:15:50	prithade@gmail.com	Pritha Dey	17-05-1978	Between 41 to 60	Ab tere bin

Fig. 5 Song recommendation based on age group and feedback

library and outside to our music library. We apply the following six songs, out of which three from our music library that is song no 2, 3 and 5 and the remaining three songs come from the outside of the music library that is song no 1, 4 and 6. We found that from the given table that only one song is for the age group 1–20, the two songs are from the age group 21–40, one song from the age group 41–60 and two songs are from the age group 61–80. Figures 6, 7 depict some screenshots of the work.

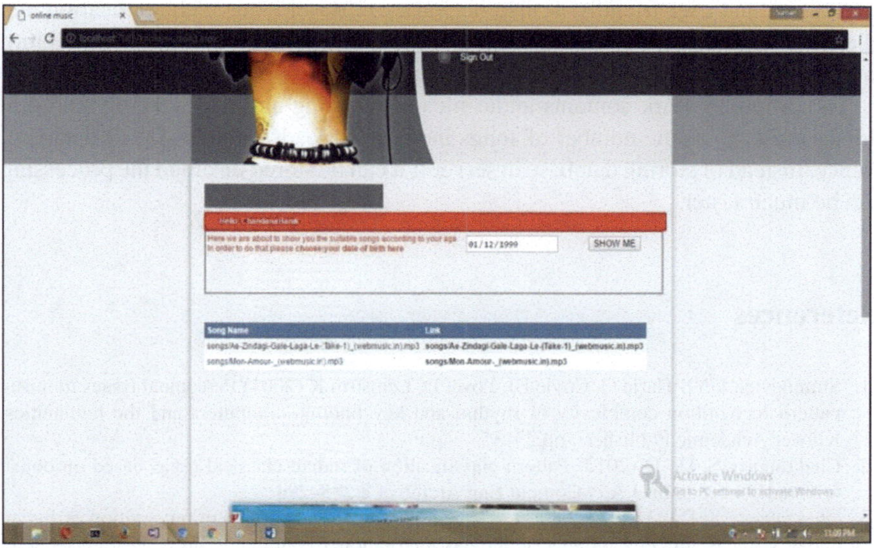

Fig. 6 Bar diagram of frequency value and their corresponding occurrences of rhythm 1

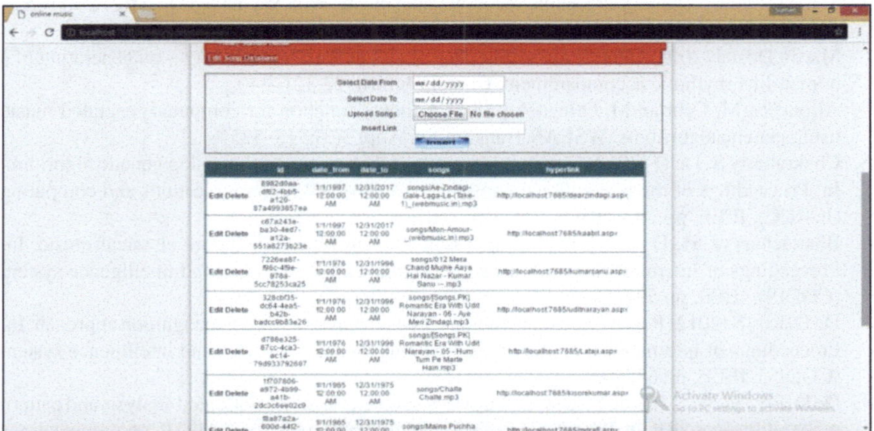

Fig. 7 Bar diagram of frequency value and their corresponding occurrences of rhythm 1

5 Conclusion

"Context-aware song recommendation system" is a Web-based application which has a great demand in the future. It is very much fast, effective and efficient. The primary objective of this innovative contribution is to calculate the age of listener and play from the song music library based on their age group. "Context-aware song recommendation system" helps to recommend the exact song of users and it is very convenient and user-friendly. It also helped us in developing Web application and client-server technology that will be a great scope in the future which will in return provide to create user preferences music library.

The proposed work contains audio files of the song and that is being stored in the database, so as the number of songs increases it leads to large size of database. Hence, instead of storing database in server if it can be stored on cloud the processing can be much faster.

References

1. Shmulevich I, Yli-Harja O, Coyle EJ, Povel D, Lemstrm K (2001) Perceptual issues in music pattern recognition complexity of rhythm and key finding. computers and the humanities. Kluwer Academic Publishers, pp 23–35
2. Chakraborty S, De D (2012) Pattern classification of indian classical ragas based on object oriented concepts. Int J Adv Comput Eng Architect 2:285–294
3. Chakraborty S, De D (2012) Object oriented classification and pattern recognition of Indian classical ragas. In: Proceedings of the 1st international conference on recent advances in information technology (RAIT). IEEE, pp 505–510
4. Gartland-Jones A, Copley P (2003) The suitability of genetic algorithms for musical composition. Contemp Music Rev 22(3):43–55
5. Matic D (2010) A genetic algorithm for composing music. Proc Yugoslav J Oper Res 20(1):157–177
6. Martin Dostal (2005) Genetic algorithms as a model of musical creativity—on generating of a human-like rhythmic accompaniment. Comput Inform 22:321–340
7. Alfonceca M, Cebrian M, Ortega A (2006) A fitness function for computer-generated music using genetic algorithms. WSEAS Trans Inf Sci Appl 3(3):518–525
8. Chakrabarty S, De D (2012) Quality measure model of music rhythm using genetic algorithm. In: Proceedings of the international conference on RADAR, communications and computing (ICRCC). IEEE, pp 203–208
9. Bhattacharyya M, De D (2012) An approach to identify that of Indian classical music. In: Proceedings of international conference of communications, devices and intelligence system (CODIS). IEEE, pp 592–595
10. De D, Roy S (2012) Polymorphism in Indian classical music: a pattern recognition approach. In: Proceedings of international conference on communications, devices and intelligence system (CODIS). IEEE, pp 632–635
11. De D, Roy S (2012) Inheritance in Indian classical music: an object-oriented analysis and pattern recognition approach. In: Proceedings of international conference on RADAR, communications and computing (ICRCC). IEEE, pp 296–301
12. Roy S, Chakrabarty S, Bhakta P, De D (2013) Modelling high performing music computing using petri nets. In: Proceedings of international conference on control, instrumentation, energy and communication. IEEE, pp 757–761

13. Roy S, Chakrabarty S, De D (2013) A framework of musical pattern recognition using petri nets. In: Emerging trends in computing and communication 2014. Springer-Link Digital Library, Submission No. 60 (accepted)
14. Chakrabarty S, Roy S, De D (2014) Pervasive diary in music rhythm education: a context-aware learning tool using genetic algorithm. In: Advanced computing, networking and informatics, vol 1. Springer International Publishing, pp 669–677
15. Chakrabarty S, Roy S, De D (2013) Automatic raga recognition using fundamental frequency range of extracted musical notes. In: Proceedings of the eight international multiconference on image and signal processing (ICISP 2014). Elsevier
16. Chakrabarty S, Gupta P, De D (2015) Behavioural Modelling of Ragas of Indian Classical Music using Unified Modelling Language", In Proceedings of the Second International Conference on Perception and Machine Intelligence (PerMIn'15), ACM Digital Library, pp 151–160
17. Chakrabarty S, Roy S, De D (2016) Time-slot based intelligent music recommender in Indian music. In: International book chapter, ISBN 13: 9781522504986, ISBN 10: 1522504982. IGI Global, USA
18. Chakrabarty S, Islam MR, De D (2017) Modelling of song pattern similarity using coefficient of variance. Int J Comput Sci Inf Seq, 388–394, ISSN 1947-5500

Author Index

H. K. D. Sarma et al. (eds.), *Trends in Communication, Cloud, and Big Data*, Lecture Notes in Networks and Systems 99, https://doi.org/10.1007/978-981-15-1624-5

The manufacturer's authorised representative in the EU is Springer Nature Customer Service Centre GmbH, Europaplatz 3, 69115 Heidelberg, Germany. If you have any concerns regarding our products, please contact ProductSafety@springernature.com

Printed and bound by CPI Group (UK) Ltd, Croydon, CR0 4YY
30/09/2026
02225574-0017